# CORPORATE RESTRUCTURING AND INDUSTRIAL RESEARCH AND DEVELOPMENT

Academy Industry Program

National Academy of Sciences
National Academy of Engineering
Institute of Medicine

NATIONAL ACADEMY PRESS
Washington, D.C. 1990

National Academy Press • 2101 Constitution Avenue, N.W. • Washington, D.C. 20418

Support for this project was provided by the Academy Industry Program.

Library of Congress Catalog Card Number 89-63979

International Standard Book Number 0-309-04186-4

Copyright © 1990 by the National Academy of Sciences

S074

Printed in the United States of America

## ACADEMY INDUSTRY PROGRAM

ALLAN R. HOFFMAN, Director
EDWARD ABRAHAMS, Senior Staff Officer
LOIS E. PERROLLE, Staff Officer
DEBORAH FAISON, Senior Program Assistant

The Academy Industry Program was established in 1983 to enhance communication between the National Research Council and industry leaders on issues related to the health of U.S. science and technology. It serves as a two-way channel of communication by providing National Research Council reports to industry and by providing a forum in which industry leaders can bring their views on important issues in science and technology to the attention of the Research Council's leadership. The program also provides financial support for institutionally initiated studies for which government funding may be inappropriate or unavailable. Over 70 companies currently participate in this program.

# Preface

If there is a cultural clash between technology and economics, it is nowhere more evident than in the dialogue about the effects of corporate restructuring on industrial research and development. The debate has grown in volume and vehemence with the recent proliferation of debt-intensive buyouts and takeovers. Members of the research community voice fears that debt service will be paid at the expense of R&D; financiers argue that restructuring improves corporate efficiency without impacting R&D expenditures.

The debate has obvious implications for public policy, since R&D is vital to the nation's ability to compete in the global marketplace, where technological advances are of great and increasing importance. Such matters are of particular concern to the National Research Council, which has a mandate to address major issues involving science and technology. Therefore in 1988 the Academy Industry Program, the Research Council's principal channel of communication with industry, began to plan a symposium on the subject.

The symposium would serve two purposes. First, by bringing together proponents of various points of view, it would help all sides learn something of the others' needs and expectations. Second, a full discussion of the issues would identify areas in which more research was needed to guide policy decisions.

These goals were largely achieved during the symposium, which took place on October 11 and 12, 1989, at the National Academy of Sciences in Washington, D.C. The speakers represented a range of opinions from government, Wall Street, industry and academia. Most in the audience represented corporate members of the Academy Industry Program—chief executives and vice presidents for research and development. Some academics, government officials and media representatives also attended.

This report is a transcript of the speakers' remarks and of the question-and-answer sessions. It also includes, as appendixes, a background paper prepared by Kenneth Flamm of The Brookings Institution (provided to all participants before the symposium convened) and a list of attendees. A summary of the discussion can be found in the concluding remarks of Stuart Eisenstat, the symposium moderator.

# Contents

# CORPORATE RESTRUCTURING AND INDUSTRIAL RESEARCH AND DEVELOPMENT

# Evening Session
# October 11, 1989

DR. PRESS: Good evening, ladies and gentlemen.

Thank you for joining us to discuss this important topic. As everyone here this evening realizes, the health—or distress, as some may see it—of U.S. industrial R&D today is going to affect American competitiveness in the future. With the wave of corporate restructuring that has swept over the American corporate scene, there is a sense of timeliness in this symposium, and I hope that with your help we will clarify some of the issues regarding its impact, particularly on research and development.

But rather than introduce the subject of the symposium, which I will leave to tonight's speaker, who has studied the issue in depth and who, I am sure, will have many provocative points to make, I would like to say a few words about the National Research Council's Academy Industry Program.

As you know, the National Research Council is the operating arm of the National Academy of Sciences, the National Academy of Engineering and the Institute of Medicine. As most of you probably know, most of the National Research Council's work is generated by requests from the federal government.

Back in 1983, we in the National Academy of Sciences, the Academy of Engineering and the Institute of Medicine decided to expand our outreach efforts, beyond our traditional constituencies in government and academia, to industry, particularly those in corporations, whose products result from advanced research and development.

We therefore established the Academy Industry Program, which is the National Research Council's official liaison to American industry. In

1

addition to putting on programs like the one tonight and tomorrow, the AIP regularly disseminates relevant NRC studies to corporate executives and keeps them informed generally about our activities.

In addition, the program receives support from approximately 70 member companies each year, which together with support we receive from philanthropic foundations allows us to initiate studies on our own and reduce our dependence on the federal government. That money enables us to ask questions and raise issues, perhaps, before they appear on the federal radar screens.

So, I welcome you to the Academy complex and to the Academy Industry Program, in particular those guests who may not have been here before.

Before introducing tonight's speaker, I want to take this opportunity to acknowledge the participation of Mac Booth, President and CEO of Polaroid, in this symposium. Polaroid is and has been one of the mainstays of the Academy Industry Program and we appreciate that.

I also want to publicly thank Stu Eizenstat, who will moderate the symposium tomorrow, for his help in putting this all together.

We are particularly fortunate tonight to have as our keynote speaker, Joseph Grundfest. Mr. Grundfest has been a Commissioner on the Securities and Exchange Commission (SEC) since 1985. He has shared responsibility for regulating this nation's securities markets during what has obviously been their most tumultuous period in recent memory, with the crash, the insider trading scandals and the growth of sophisticated institutional investment instruments.

His background helped prepare him for this position. He is both an attorney and an economist, and prior to becoming a Commissioner at the SEC, he was a senior economist on the President's Council of Economic Advisors, with responsibility for legal and regulatory matters. He has practiced law and has worked at both the Rand Corporation and the Brookings Institution.

Mr. Grundfest is author or co-author of numerous research reports and publications. His works deal with a range of topics, including competition for corporate control, insider trading, international securities regulation, regulation of markets subject to kick-back schemes, the economics and regulation of broadcasting and the role of citizen participation in administrative proceedings.

Mr. Grundfest's education is equally varied. He received a BA in economics from Yale and a master's degree in econometrics from the London School of Economics. In addition, he earned a law degree from Stanford in 1978. Among his honors has been a National Science Foundation Fellowship and I know those are highly competitive.

When Mr. Grundfest leaves the Securities and Exchange Commission sometime next year, he will go to Stanford, where he will teach at the law school. Also, next year, Little, Brown will publish a collection of his essays and speeches under the title, *Perestroika on Wall Street.*

With that, it gives me great pleasure to introduce him to you. Mr. Grundfest.

The following is the prepared text of Commissioner Grundfest's talk. The views expressed are those of Commissioner Grundfest and do not necessarily represent those of the U.S. Securities and Exchange Commission, of other commissioners, or of the Commission's staff.

It is an honor and privilege to be invited to address this distinguished gathering at the National Academy of Sciences. The privilege is particularly great because it affords an opportunity to consider a topic of substantial national concern: the relationship between America's research and development efforts and the growing wave of restructuring activity now sweeping our corporate sector.

It probably comes as little surprise that hordes of critics stand ready to condemn corporate restructuring in the United States as bad for workers, harmful to local communities, damaging to U.S. international competitiveness, and threatening to the financial stability of the U.S. economy.[1] The most damaging allegation against corporate restructuring may, however, be the charge that it stifles corporate R&D and forces management to adopt irrationally myopic strategies.[2] To my mind, this allegation is the most serious potential indictment because, if true, it suggests that restructuring is eroding America's industry at the point where it is most vulnerable: at its knowledge base.

The reality of today's marketplace is that a firm's knowledge capital measured in terms of its know-how, technical expertise, trade secrets, patents, productive processes, and accumulated R&D efforts is often far more important than its physical capital measured in terms of bricks and mortar, lathes, delivery trucks, or power plants.[3] More and more, capital

---

[1] See, e.g., M. Lipton, *Corporate Governance in the Age of Finance Corporatism,* 136 U. Pa. L. Rev. 1 (1987); L. Lowenstein, *Management Buyouts,* 85 Colum. L. Rev. 730 (1985); *LBO's: Friend or Foe of Industrial Research?,* 31 Research & Development 13 (April 1989); *What Are LBO's Doing to R&D?,* Chemical Week, Feb. 15, 1989, at 26; Thurow, *U.S. Can't Compete if Finance Continues as the Master of Industry,* L.A. Times, Nov. 17, 1985, Pt. 5, at 3.

[2] See, e.g., J. Stein, *Takeover Threats and Managerial Myopia,* 96 J. Polit. Econ. 61 (1988); P. Drucker, *A Crisis of Capitalism,* Wall St. J., Sept. 30, 1986, at 32, col. 3; *"Long-Term" Bandwagon Hot,* Pensions & Investments Age, May 1, 1989. ("Business executives have argued that short-term pressures placed on many mangers force . . . [executives] to take a short-term view.")

[3] See, e.g., B. Hall and F. Hayoshi, *Research and Development as an Investment,* Nat. Bur. Econ. Res. Working Paper No. 2973, May 1989, at 2, 33.

exists as software that is invisible to the eye rather than as hardware that we can touch or feel. This "knowledge capital" is not easily inventoried by accountants or valued by appraisers. In many situations, it is not even easily described. Yet, while knowledge capital cannot easily be measured, without it our economy surely will not thrive.

Unfortunately, when it comes to evaluating our stock of knowledge capital the indications for the U.S. economy are quite depressing. At the most elementary level, our school children do not read, write, or add as well as school children in the countries that are our strongest economic competitors.[4] The test scores of entering college freshmen are in a long-term decline, having peaked in 1963 and having recovered only slightly from a 1980 trough.[5] The adult literacy rate is a national disgrace.[6] The educational situation in our country is in such shambles that many companies have to provide remedial literacy and arithmetic training on the job just so they can have a work force competent to operate an efficient production process.[7]

The time has passed for the production line worker who knows only how to turn a wrench or drive a screw. Technology will not create high-paying, high-quality jobs for a nation of illiterates who cannot add. Until we turn this situation around, and start educating a work force smart enough to master the demands of modern production technology, it is hard to understand how we can expect to make progress against competitors who can read operating manuals that we can't and perform statistical analyses that we don't understand. Viewed from the shop floor, that's the cold reality of the world we face today.

Viewed from the corporate boardroom, our comparative situation may not be much better. Although corporate spending on R&D climbed steadily at an inflation-adjusted rate of 5.8 percent per year in the decade preceding 1986, aggregate corporate spending on R&D has slowed substantially since

---

[4] See A. McLaughlin, *Education and Work: The Missing Link*, N.Y. Times, Sept. 25, 1989, at A19, col. 1.; Vagelos, *The Sorry State of Science Education*, Scientific American, Oct. 1989 at 128.

[5] See L. Feinberg, *Student's Scores Drop in Test of Verbal Skills*, Wash. Post, Sept. 12, 1989, at A18. *See also Apptitude Test Scores Drop for Women and Minorities*, L.A. Times, Sept. 11, 1989, at A2, col. 3.

[6] See A. McLauglin, *Education and Work, supra* n. 4, ("Between 20 million and 40 million adults today have literacy problems"); S. Knell, *An Investment in Human Betterment—Adult Literacy*, Chi. Tribune, Sept. 18, 1989, at C13 ("There are an estimated 20 million to 30 million [illiterate] adults in the United States"); M. Spencer, *Why Won't Johnny Learn? Look in the Mirror*, L.A. Times, Sept. 9, 1989, at Part 9, page 1, col. 1 ("adult illiterary [is] running at a 20% rate").

[7] See J. Berger, *Skills v. Jobs: The Classroom Mismatch*, N.Y. Times, Sept. 26, 1989, at A1, col. 2; C. Skrzycki; *The Company as Educator: Firms Teach Workers to Read, Write*, Wash. Post, Sept. 22, 1989, at G1.

then.[8] When these figures are adjusted to reflect R&D/sales ratios and inflation, the recent slowdown in aggregate R&D expenditure may not appear quite as foreboding.[9] However, once we compare U.S. research expenditures with our Japanese and West German competitors, the picture turns dim again.

Nondefense corporate spending on R&D in the United States stood at about 1.9 percent of gross national product (GNP) as of year-end 1987.[10] In Japan, the comparable measure of R&D spending ran at 2.7 percent of GNP and in West Germany, it stood at 2.6 percent of GNP.[11] Most ominous, perhaps, is the fact that Japanese commercial R&D expenditures are growing at a far faster rate than comparable U.S. expenditures.[12] Thus, not only are we falling behind at the most elementary levels of our knowledge base, but if the trajectory of R&D expenditures as a percentage of GNP is a harbinger of future trends, then it appears that we may also be falling behind at the most advanced levels of our knowledge base as well.

Simply put, we are in deep trouble. If we don't invest in our knowledge base—both at the top and at the bottom—we are going to lose out in the battle for international economic competitiveness. The question is not if we will lose out; the only question is when.

Faced with this serious predicament, wouldn't it be wonderful if we could find a quick, simple, and popular cure for at least part of this

[8] See, e.g., Testimony of Erich Bloch, Director, National Science Foundation, Before the House Ways and Means Committee, March 14, 1989, at 2 (hereinafter cited as "Bloch Testimony"); *Business Talks a Better R&D Game Than It Plays*, Bus. Wk., Aug. 21, 1989, at 20 (economists estimate that real outlays on research and development will increase by less than 1% in 1989, compared with 1.3% in 1988 and 3.7% in 1987); R. Cassidy, *Research Funding for 1989 Won't Even Reach $131 Billion*, Research & Development, Jan. 1989, at 47; R. Winter, *Research Spending in U.S. to Slow in 1989*, Wall St. J., Dec. 21, 1988, at B3, col. 1.

[9] K. Flamm, *Industrial Research and Corporate Restructuring: An Overview of Some Issues*, September 1989, at 4-5. ("In the face of a declining sales base in 1985 and 1986, the relative size of the research efforts of American R&D-performing companies increased. When sales picked up in 1987, the interest of these companies' research efforts decreased.")

[10] U.S. Department of Commerce, *Statistical Abstract of the United States*, 1989 at 578, Table 973 ("Statistical Abstract"); *Missed Opportunities: R&D—A Bigger Push in Japan*, Wall St. J., Nov. 14, 1988, at R21, col. 3. *See also* Clark & Malabre, *Slow Rise in Outlays for Research Imperils U.S. Competitive Edge*, Wall St. J., Nov. 10, 1988, at A1, col. 6. For purposes of the present analysis I am focussing on nondefense R&D because it is unlikely that restructuring activity has a meaningful effect on defense R&D spending levels. In addition, nondefense R&D expenditures are more directly related to the economy's international economic competitiveness than military R&D expenditures. See, e.g., Reich, *The Quiet Path to Technological Pre-eminence*, Scientific American, Oct. 1989, 41, 44. ("Several factors impede technology transfer from military to commercial applications . . . military R&D has become an inefficient means of generating commerical spin-offs. . . .")

[11] Statistical Abstract, *supra* n. 10; *Missed Opportunities*, *supra* n. 10.

[12] *Missed Opportunities*, *supra* n. 10, at R22.

problem? Wouldn't it be wonderful if we could establish that corporate restructuring has become a millstone around the neck of America's R&D efforts? Wouldn't it be wonderful if we could just stop all this restructuring and thereby restore at least part of America's R&D vitality?

Yes, it would be wonderful, but it wouldn't be true. The best available evidence suggests that corporate restructuring has relatively little to do with our declining international position in the R&D race. Thus, even if we placed substantial constraints on takeovers, leveraged buyouts, spinoffs, stock buybacks, leveraged recapitalizations, and other forms of corporate restructuring, I doubt that we would accomplish much, if anything, in terms of restoring the vitality of America's R&D efforts. The facts that lead me to this conclusion are not pretty, but with your indulgence I'd like to review them in some detail.

## THE COST OF CAPITAL

If corporate restructuring is not a major cause of the relative decline in U.S. R&D activity then what is? One need not search hard or long for the answer to this question because the primary culprit is quite clear: it is the cost of capital.

The evidence is overwhelming that the cost of capital for R&D projects in the United States is far higher than it is in Japan or West Germany. Because our capital costs are higher, it is more expensive to conduct R&D in the United States. Moreover, because our capital costs are higher, the projects we conduct must have faster payoff periods, and we cannot afford to undertake projects that are as risky as some projects conducted in Japan or West Germany.

Just how much higher are our capital costs than Japan's or West Germany's? Recent estimates by staff of the Federal Reserve Bank of New York are truly frightening.[13] After adjusting for inflation, tax rates, and other factors, the New York Fed study suggests that in 1988, the average annual effective cost of capital in the United States for a benchmark R&D project was 20.3 percent. The cost of capital in West Germany for the same benchmark project was 14.8 percent, and in Japan it was only 8.7 percent.[14] Thus, capital for R&D purposes is now more than twice as expensive in the United States as in Japan. Unfortunately, the conclusion that U.S. capital costs are higher than foreign capital costs is borne out in several recent studies.[15]

---

[13]R. N. McCauley and S. A. Zimmer, *Explaining International Differences in the Cost of Capital*, Federal Reserve Bank of New York, Quarterly Review 7, 16 (Summer 1989).

[14]*Id.*

[15]B. D. Bernheim and J. Shoven, *Taxation and the Cost of Capital: An International Comparison* in C. E. Walker and N. A. Bloomfield, *The Consumption Tax: A Better Alternative? (1987)*,

These cost of capital figures are consistent with calculations of required breakeven periods for capital projects in the United States and Japan. For example, a recent Stanford University study suggests that capital costs in the United States imply an average breakeven period for new investment projects of 5.7 years. In contrast, lower capital costs in Japan push the Japanese breakeven period out to 10.3 years.[16] Under these circumstances, an eight-year research project that seems perfectly reasonable to a Japanese manager may be totally out of the question for a U.S. manager—not because the American lacks the wisdom, vision, or will, but simply because Americans can't rationally bear the cost of capital.

The implications of these high capital costs were well explained by Carl Ledbetter, formerly the president of ETA, the supercomputer division of Control Data that was recently shut down in response to competition from Hitachi, Fujitsu, NEC, and others.[17] As Mr. Ledbetter put it, "If our capital costs had been lower, ETA could have survived."[18] Touché.

If our capital costs were half of what they are today, our R&D efforts would mushroom. With lower capital costs we too could afford to be much more patient in waiting for R&D efforts to pay off; and we too could afford to take risks that are just too far-fetched when you're facing a hurdle rate of return of 20 percent per year.

The key then is to identify the cause of high capital costs in the U.S. markets and strike at those economic factors. The most recent evidence suggests that our capital costs are higher than Japan's predominantly because our savings rate is lower and because we have less macroeconomic stability as reflected in the volatility of price levels and interest rates.[19]

---

at 78. Bernheim and Shoven's estimates indicate that the cost of capital calculated at the average interest and inflation rates for the 1980s, using 1985 tax codes, was 5.48 percent in the United States, 4.39 percent in West Germany, and only 2.76 percent in Japan. In addition, G. Hatsopoulis and S. Brooks, *"The Gap in the Cost of Capital: Causes, Effects, and Remedies,"* in R. Landau and D. Jorgenson, *Technology and Economics Policy* (1986) estimate that the cost of capital in the United States is almost three times higher than in Japan. For other estimates *see* A. Ando and A. Auerback, *The Cost of Capital in the United States and Japan*, 2 J. Japanese and Int. Economics 134 (1988); A. Ando and A. Auerbach, "The Corporate Cost of Capital in Japan and the United States: A Comparison," in J. Shoven, *Government Policy Towards Industry in the United States and Japan* (1988); Corcoran and Wallich, *The Analytical Economist: The Cost of Capital*, Scientific American, Oct. 1989, at 79.

[16] Bernheim and Shoven, *supra* n. 15. See also L. Richman, *How Capital Costs Cripple America*, Fortune, August 14, 1989, at 50.

[17] Richman, *supra* n. 18, at 50.

[18] Richman, *supra* n. 16, at 50.

[19] See, e.g., McCauley and Zimmer, *supra* n. 13. ("Higher household savings in Japan and Germany and more sucessful policies for maintaining stable growth in Japan and stable prices in Germany have opended up the [capital cost] gap.")

There is little or no support for the idea that corporate restructuring is a meaningful cause of high capital costs in the United States.

Thus, even if we shut down all corporate restructuring activity we would still have a substantially higher cost of capital and we would still find ourselves falling behind Japan in the international R&D race. Moreover, if one wants to attribute causality, I think the better argument is that the high cost of capital in the United States causes both a decline in relative R&D intensity and an increase in corporate restructuring activity.[20] Restructuring and R&D are thus codetermined by interest rates, and restructuring does not, in and of itself, determine R&D intensity. Put another way, increased restructuring activity does not cause a decline in R&D activity any more than an increase in R&D activity would cause a decline in restructuring.

Addressing this problem will not be easy. To bring down the cost of capital we must increase our domestic savings rate and restore macroeconomic stability over a substantial period of years. That's nowhere near as exciting, dramatic, or simple as putting an end to corporate restructuring, but I'm afraid that this painful prescription is the only one likely to do meaningful good.

## THE ARGUMENT BEYOND CAPITAL COSTS

Having argued that capital costs explain the lion's share of the R&D problem facing American industry, I could end right here and suggest that we stop blaming corporate restructuring for something that isn't its fault. While this approach has the virtue of being concise, it suffers from the vice of overlooking much additional evidence exploring the relationship between restructuring and R&D. To develop this perspective more fully, let's proceed on the obviously unrealistic assumption that capital costs have nothing to do with the R&D decision, and let's explore the evidence that focuses narrowly on the relationship between restructuring, stock market valuation, R&D activity, and the alleged myopia of America's capital markets.

## THE SECTORAL INCIDENCE OF RESTRUCTURING ACTIVITY

If corporate restructuring activity truly has a major impact on aggregate R&D expenditures, we would expect to find that corporate restructuring occurs with some frequency in industries that engage in significant amounts

---

[20]High capital costs imply a greater scarcity value for capital. This higher price of capital suggests that the market will be less tolerant of managements that fail to earn adequate returns. Corporations that accumulate "free cash flow," *i.e.*, cash flow that could be reinvested outside the corporation more profitably than it can be reinvested in the corporation, are wasting valuable capital and are potential subjects for restructuring. See, e.g., Jensen, *The Agency Cost of Free Cash Flow*, 76 Amer. Econ. Rev. 490 (May 1986).

of R&D. The data, however, fail to support this hypothesis. As explained by Professor Lawrence Summers of Harvard University, Governor Dukakis's chief economic adviser during his recent presidential campaign, "[m]ost LBOs occur in mature industries that do not spend a lot on research, so there is not yet much evidence to support claims that R&D is severely cramped by LBOs."[21] Indeed, most restructuring takes place in industries that are not R&D intensive, so the point extends beyond simple LBO transactions.[22]

This observation can be made much more simply: Bloomingdales didn't do much R&D before its takeover and it hasn't done much R&D since. The same holds true for takeovers, buyouts, and restructurings involving companies like Beatrice, United Airlines, Stop N' Shop, Allied Stores, Avin, Storer Cable, Columbia Pictures, Burlington Industries, and hundreds of others engaged in non-R&D intense lines of business. Indeed, an examination of all takeovers in 1988 suggests that more than 75 percent of the dollar value of merger and acquisition (M&A) transactions occurred in industries such as retailing, food products, broadcasting, and insurance—industries in which R&D is not perceived as a major competitive factor.[23]

Measures aimed at hobbling takeovers and buyouts in the name of protecting America's R&D effort are therefore clearly overbroad because they would impose substantial restrictions on transactions that have nothing to do with R&D. Indeed, Professor Frank Lichtenberg of Columbia University has suggested that managers and administrators who typically lose jobs following restructurings may be raising false alarms about the general effects of takeovers on R&D.[24] Thus, the R&D argument may make good public relations in the campaign against restructuring, but it is totally irrelevant to the large majority of restructurings to which the argument might be applied.

## R&D INTENSITY OF RESTRUCTURING TARGETS

But suppose we pursue the argument further and eliminate from consideration companies like Bloomingdales where R&D isn't a meaningful issue. What then do we find? This question can be posed in two ways. First, what difference is there in the R&D intensity of restructuring targets prior

---

[21] L. Summers, *LBO Debt and Taxes*, Across the Board, Vol. XXVI, No. 4, April 1989, at 53, 54.

[22] See, e.g., B. Hall, *Leveraged Buyouts, Corporate Debt, and R&D Investment: Is There Any Connection?* (work in progress, version of Sept. 1989) at 6.

[23] *1988 Profile; Merger Activity by Industry Area*, Mergers and Acquisitions, at 54 (May/June 1989).

[24] Testimony of Frank R. Lichtenberg in hearings on "Corporate Restructuring and Its Effects on R&D" before the Science, Research, and Technology Subcommittee of the House Committee on Science, Space, and Technology, July 13, 1989, at 2.

to restructuring activity—do they conduct more or less R&D than their industry peers? Second, what difference is there in the R&D intensity of firms that have been restructured—do they conduct more or less R&D after the restructuring than before? Posed either way, the answer to the question is fascinating and again fails to support the argument that restructuring is a primary factor hobbling U.S. R&D activity.

Let's look first at the data describing the prerestructuring R&D intensity of firms that conduct a meaningful amount of R&D. A recent study by economists at the Securities and Exchange Commission's (SEC) Office of Economic Analysis found that on average, "takeover targets undertake less R&D than non-targets in the same industry."[25] This result is hardly novel and reaffirms earlier consistent findings at the SEC,[26] at Harvard,[27] and at the National Bureau of Economic Research that also show that much of the takeover activity in the U.S. market was directed "toward firms and industries that were relatively less R&D intensive and had a weaker technological base."[28]

The implications of this result are quite significant. If companies that engage in above average amounts of R&D are setting themselves up to be targets of restructuring activity then we should find that, within industries, takeover activity is targeted at highly R&D-intensive firms. Instead, the data support exactly the opposite conclusion and it appears that industry laggards who fail to do as much R&D as their counterparts are more likely to be takeover targets than the industry's takeover leaders. Thus, the image of America's R&D leaders as being under the restructuring gun appears to be at odds with at least one large and inconsistent fact: The companies feeling the heat from restructuring activity appear to be those that have failed to invest as much in R&D as their industry counterparts—not the other way around.

## R&D CHANGES AFTER RESTRUCTURING

Having established that the bulk of restructuring activity occurs in industry segments that are not R&D intensive, and that the targets of restructuring activity tend to do less R&D than their industry peers, a

[25] L. Meulbroek et al., *Takeover Threats and Research & Development: Testing Stein's Model of Managerial Myopia*, at 6 (1989), J. Polit. Econ. (forthcoming).

[26] Securities and Exchange Commission, Office of the Chief Economist, *Institutional Ownership, Tender Offers, and Long Term Investment*, 1985.

[27] S. Addanki, *Innovation and Mergers in U.S. Manufacturing Firms: A First Look*, Department of Economics, Harvard University, 1985 (*cited in* Hall, *supra* n. 22).

[28] B. Hall, *The Effects of Takeover Activity on Corporate Research and Development* at 93, *cited in* A. Auerbach, *Corporate Takeovers: Causes and Consequences* (1988).

meaningful question nonetheless remains: What happens to R&D at those companies that actually do R&D and that are subject to restructuring?

At this point we encounter a hailstorm of anecdotal evidence suggesting that R&D expenditures are viciously slashed in the wake of corporate restructuring efforts.[29] For example, one economist claims that "one of the things that gets squeezed [in a restructuring] is R&D, because that's an investment in the future. . . . Whatever costs are postponable are likely to go by the boards."[30] Parsing the evidence on this score is quite an interesting exercise because even if there is a postrestructuring decline in R&D/sales ratios, that decline could reflect greater efficiency resulting from economies of scale. A decline in R&D expenditures could also reflect a decision to kill R&D projects that have become white elephants. However, before pursuing these avenues of inquiry, let's take a step back and explore the evidence about postrestructuring R&D expenditures. Do R&D expenditures increase, decrease, or stay about the same following a corporate restructuring?

The evidence on this score is more mixed than for the other points I have discussed. However, based on current research, I would characterize the data as either supporting the "no difference" conclusion or as too uncertain to support any conclusion. Lichtenberg and Siegel, for example, found that "the average R&D intensity of firms involved in LBO's increased at least as much from 1978 to 1986 as did the average R&D intensity of all firms responding to the [National Science Foundation] NSF/Census survey of industrial R&D."[31] In earlier research the same authors found that R&D employment does not change following restructuring, even though there is a substantial decline in nonproduction employment, that is, of managers and administrators who work at corporate headquarters.[32]

Similarly, Bronwyn Hall examined all takeovers of publicly traded manufacturing firms between 1976 and 1986 and concluded that the data "provide very little evidence that acquisitions cause a reduction in R&D spending." In the aggregate the firms involved in mergers were in no way different in their pre- and postmerger R&D performance from those not

---

[29] See, e.g., Statement of Dr. Julie Fox Gorte, project director, Office of Technology Assessment, Before the Subcommittee on Science, Research, and Technology, Committee on Science, Space, and Technology, U.S. House of Representatives, July 13, 1989, at 5-10.

[30] C. Skrzycki, *Impact on R&D is Newest Worry About LBO's*, Wash. Post, Dec. 18, 1988, at H1, col. 3 (quoting Walter Adams, professor of economics at Michigan State University). C. Skrzycki, *Impact on R&D is Newest Worry About LBO's*, Wash. Post, Dec. 18, 1988, at H1, col. 3 (quoting Walter Adams, Professor of Economics at Michigan State University).

[31] F. Lichtenberg and D. Siegel, *The Effects of Leveraged Buyouts on Productivity and Related Aspects of Firm Behavior*, Nat. Bur. Econ. Res., Working Paper No. 3022, June 1989, summary.

[32] F. Lichtenberg and D. Siegel, *The Effects of Takeovers on the Employment and Wages of Central Office and Other Personnel*, Nat. Bur. Econ. Res., Working Paper No. 2895, March 1989.

so involved. At the individual industry level, however, the results were too imprecisely measured to draw solid conclusions.[33]

Abbie Smith, in a study of 58 management buyouts (MBOs) completed between 1977 and 1986, finds a "substantial increase in profitability following the MBO."[34] She concludes, however, that these increased profits are apparently not due to "pervasive cutbacks in 'discretionary expenditures' such as maintenance and repairs, advertising, or R&D which might lead to a longer run decline in cash flows."[35] In particular, Smith finds that the "median ratio of R&D expense to sales increases from .012 in the year preceding the MBO to .018 in the year following the MBO, with a median change of 0.00 for the seven firms with available data."[36]

Moreover, there is anecdotal evidence that R&D expenditures may actually increase in some situations following a takeover or restructuring. For example, after Hoechst's purchase of Celanese Corporation, R&D spending increased by 10 percent annually.[37] Data prepared by Kohlberg, Kravis & Roberts, America's leading leveraged buyout firm, also confirms that R&D expenditures decline prior to leveraged buyouts and suggests that KKR, at least, budgets for aggregate increases in R&D expenditure.[38]

On the other side of the ledger, however, stands a recent NSF study that examined R&D expenditures at the 200 largest R&D-performing firms in the United States.[39] These firms account for almost 90 percent of industrial R&D spending in the United States. Within this sample, the

---

[33] Hall, *supra* n. 28, at 93.

[34] A. Smith, *Corporate Ownership Structure and Performance: The Case of Management Buyouts,*" Univ. Chi. June 1989, at 1.

[35] *Id.* at 2.

[36] *Id.* at 24.

[37] Testimony of Dr. Julie Fox Gorte, *supra* n. 29, at 93.

[38] Kohlberg, Kravis, and Roberts, *Presentation on Leveraged Buyouts* at 8-1 (Jan. 1989).

[39] National Science Foundation, *Corporate Mergers Implicated in Slowed Industrial R&D Spending,* Washington, March 1989 ("NSF Study"); Testimony of Erich Bloch, director, National Science Foundation, House Ways and Means Committee, March 14, 1989 ("Bloch Testimony").

I exclude from consideration the findings of the Ravenscraft-Scherer studies, which, based on 1977 data, find that "lines of business originating from mergers had significantly lower company-financed R&D to sales ratios" than similar companies without a merger history. D. Ravenscraft and F. M. Scherer, *The Long Run Performance of Mergers and Takeovers,* at 44, in M. L. Wiedenbaum and K. W. Chilton, *Public Policy Toward Corporate Takeovers* (1988). *See also* D. Ravenscraft and F. M. Scherer, *Mergers, Sell-Offs & Economic Efficiency* (1987). These findings do not shed much light on the current controversy because the data result primarily from a conglomerate restructuring wave that is substantially different from current restructuring phenomena. Moreover, the low R&D intensity found at merged plants may simply reflect the finding that industry laggards in R&D are more likely to be involved in restructuring and may not support the hypothesis that restructuring causes a reduction in R&D efforts.

NSF identified 33 firms that were merged into 16 companies, as well as 8 freestanding firms that were involved in LBOs.

Interestingly, in the NSF sample the average R&D outlay for the 16 merged firms was $575 million per year, whereas the average R&D outlay for the 8 LBOs was only $75 million per year.[40] This statistic supports the view that LBO transactions tend to concentrate in mature, stable industries with "reliable and stable cash flows necessary to amortize the acquisition debt. . . . LBOs in research industries are rare."[41] The market thus does not want to load debt onto R&D-intensive firms because, among other reasons, debt capital is relatively more expensive than equity for R&D applications.[42]

The NSF study found, however, that these 24 restructured companies reported a 5.3 percent reduction in R&D spending while all other companies in the NSF sample reported a 5.4 percent increase.[43] All 8 of the LBO firms reduced their R&D expenditures, and the aggregate decline in R&D expenditures at these firms was 12.8 percent.[44] Indeed, even in the chemical and pharmaceutical industry, where merged companies reported a 5.4 percent increase in spending, the rest of the industry reported a 9.8 percent increase, suggesting that the merged firms were not keeping pace with industry R&D developments after restructuring.[45]

Why do the NSF results differ from the other findings that indicate no statistically significant change in R&D following restructuring? One

---

In addition, shortly before the date of delivery of this address I received a copy of Hitt, Hoskisson, Ireland, and Harrison, *Acquisitive Growth Strategy and Relative R&D Intensity: The Effects of Leverage, Diversification, and Size* (Texas A&M, Baylor, and Clemson universities, May 1989) ("Hitt Study"). The Hitt Study examined 191 mergers of publicly traded firms conducted between 1970 and 1986. Conglomerate acquisitions were the dominant form of transaction. *Id.* at Table 1. Conglomerate acquisitions are, however, quite different from current restructuring efforts, and many of the factors that caused conglomerate acquisitions to fail provide incentives for current restructuring activity. See also, Porter, *From Competitive Advantage to Corporate Strategy*, 64 Harvard Bus. Rev. 43 (1987) (documenting the failure of conglomerate acquisitions by large corporations during the period 1950-1986). In particular, spin-offs, bust-ups, and downsizings are all aimed at undoing many of the inefficiencies associated with the conglomerate form. The Hitt Study found that in conglomerate acquisitions, "acquisitive growth, leverage, diversification and size were negatively related to R&D intensity, adjusted for industry R&D intensity." Hitt Study, Abstract.

[40] Bloch Testimony at 3-4. (Derived from an NSF Study statistic reporting that the 16 merged firms spent $9.2 billion on R&D and the 8 LBOs spent $600 million.)

[41] Merrill Lynch, *Leveraged Buyouts in Perspective*, at 7 (March 1989).

[42] See, e.g., B. Hall, *How Is R&D Financed?*, Univ. Calif., Berkeley, 1989.

[43] NSF Report at 3. These changes are measured in constant dollars over the period 1986 to 1987.

[44] *Id.* at 5.

[45] *Id.*

answer is that comparing the NSF study to the other studies is a bit like comparing apples to oranges. Aside from the obvious fact that samples and time periods differ, it should be noted that the other studies measure changes in R&D intensity, typically expressed in terms of an R&D/sales ratio, while the NSF study measures aggregate R&D expenditures. One of the consequences of restructuring is typically a downsizing of the firm's scale as it focuses on more profitable market niches. Thus, in order to compare the NSF results with prior research, it may be necessary to recalculate the NSF findings in terms of research intensity.[46]

A second potential explanation of the reduction in aggregate expenditures is, as the NSF study itself notes, that "firms may simply be eliminating duplication and inefficiency within their R&D programs."[47] Here, there is at least some anecdotal evidence supporting the view that postrestructuring reductions in aggregate outlays do not necessarily imply a weakened R&D initiative.

For example, in 1986 Exxon spun off its Reliance Electric division to a management-led LBO. Management recognized that Reliance had been spending $30 million a year on overlapping research efforts and proceeded to rationalize its expenditures so that it didn't "have three people working on the same thing."[48] This rationalization chopped R&D expenditures by $25 million in 1987, or 17% of the firm's total budget.[49] At the same time, however, Reliance increased spending on related productivity tools, such as computer software and custom chips, which may not show up in R&D statistics. As one of Reliance's vice presidents explains, "We are executing projects faster, more efficiently, and experiencing less waste because we have to. Our livelihood depends on it. We're now competitive after the LBO, no question about it."[50]

Restructuring can also cause changes in the focus of research even if it does not change aggregate expenditures. Japan has built its enviable commerical position not by concentrating on basic research but by emphasizing superior commercialization. The world's videorecorder, semiconductor, and television markets are all built on basic U.S. research and Japanese commercialization. Given a choice between being a hero for doing profitless basic research or, at the margin, moving resources more vigorously into profitable commercialization, it may well make sense to reallocate resources toward the commercialization end of the R&D spectrum.

---

[46] *Accord*, B. Hall, *Leveraged Buyouts, Corporate Debt, and R&D Investment: Is There Any Connection?* Univ. of Calif., Berkeley (Work in Progress, Version of Sept. 1989, at 7).

[47] NSF Study at 4.

[48] A. Ramirez, *What LBOs Really Do to R&D Spending*, Fortune, March 13, 1989, at 98.

[49] *Id.*

[50] *Id.* (quoting Peter Tsivitse, vice president, Reliance Electric).

Several companies have recently reached just that decision. Xerox's Palo Alto Research Center spawned several successful innovations that have failed to earn Xerox a fraction of what they could have. Xerox's inability to capitalize on its development of Ethernmet, of the laser printer, and of the icon-based operating system popularized by Apple Computer is perhaps the most poignant example of a company that did its "R" brilliantly only to watch the profits slip away as a result of poor "D."[51] To minimize the chance of this happening again, Xerox has taken strong steps to assure that the research for which it is paying develops products that return value to the corporation and to its shareholders.

Success in the marketplace thus requires a balance between "R" on the one hand and "D" on the other. Without great research, there is nothing to commercialize. Without great commercialization, you never earn the fruits of your research.

It is my subjective assessment, based on recent developments in Japan and elsewhere, that a shift in our emphasis toward commercialization might be the most profitable change American industry could make in allocating its R&D. If that is a direction in which restructuring is driving America's R&D efforts, it's hard to conclude that it's all for the bad.[52]

Evidence of reduced or dramatically changed R&D expenditures therefore does not, in and of itself, suggest a weakening of a company's commitment to R&D. Nor does it necessarily suggest a reduction in the effectiveness of a company's R&D program. Instead, what we need to measure is how "smart" we are in spending our R&D dollars, because the elimination of a "dumb" R&D dollar resulting from waste, duplication, or bad planning means something quite different than the elimination of a "smart" R&D dollar that reflects a potentially profitable gamble on the scientific unknown.

## THE STOCK MARKET AND R&D EXPENDITURES

But how do we tell "smart" R&D from "dumb" R&D? The short answer is that there is no easy answer. Research and development is a gamble on the unknown. It will always be impossible to know whether two

---

[51] Pitta, *Bean Counters Invade Ivory Tower*, Forbes, Sept. 18, 1989, at 198.

[52] *Accord*, Reich, *The Quiet Path to Technological Preeminence*, Scientific American, Oct. 1989, at 41. ("If the U.S. is to regain its technological prominence, it must improve the capacity of Americans to use technology. This quiet path back to competitiveness depends less on ambitious government R&D projects . . . than on improving the way by which technological insights—whenever they may be discovered around the globe—are transformed by American workers into high quality products.") This may be one of the few points regarding economic policy on which Reich and I agree.

guys wearing white coats in a Topeka lab will, if left alone for a decade, come up with cold fusion or superconductivity.

While there is no easy answer to this question, many critics of restructuring and of the stock market would be quick to conclude that the stock market is incapable of judging the value of R&D projects and invariably penalizes companies that are committed to substantial, long-term R&D expenditures.[53] Therefore, however one judges R&D, one should surely ignore the stock market's valuation, at least according to these critics.

But is this highly negative view of the stock market's response to R&D supported by the evidence? To pose the issue most starkly, let me begin by asking a question: Which company does the stock market value more highly, Merck, a research-intensive pharmaceutical firm with 1988 sales of $5.9 billion, or General Motors, the automotive giant with 1988 sales of $110 billion that are 19 times as large as Merck's sales? Believe it or not, as of December 31, 1988, the stock market valued Merck's stock at $26.433 billion, about $400 million more than General Motors stock, which was trading at an aggregate value of $26.027 billion.[54]

But how can that be? After all, Merck is one of the most R&D-intensive companies in one of the most R&D-intensive industries in the world. In 1988 Merck spent $669 million on R&D: that's 11.3 percent of its sales, 34.9 percent of its profits, and $15,962 per employee.[55] These expenditures are for R&D projects that are wildly expensive,[56] more likely to fail than to succeed, and certain not to yield revenues in the United States for about 8 to 10 years from inception.[57] Yet Merck's stock trades at a price-earnings ratio of 23, more than triple the multiple of 7 accorded GM's shares.[58]

If the critics are right, and if the stock market is simply too impatient or myopic to wait for the payoff from R&D, then Merck's shares should be trading at an aggregate value far below General Motor's. But Merck's shares aren't trading below General Motors, and that fact takes at least some of the wind out of the sails of market critics.

While this simple comparison of Merck and General Motors is not enough to sustain any broad hypothesis about stock market behavior, it is enough to force critics of takeovers to take pause and to reconsider some

---

[53] See, e.g., the materials cited in notes 1 and 2, *supra*.

[54] *The World's 100 Largest Public Companies*, Wall St. J., Sept. 22, 1989, at R14.

[55] *R&D Scoreboard*, Business Week, Innovation 1989, at 198.

[56] The average cost of developing a new drug (new molecular entity) through approval by the Food and Drug Administration is $125 million measured in 1986 dollars. Pharmaceutical Manufacturers Association, *Facts at a Glance*, 13 (1989).

[57] *Id.* at 15, 18.

[58] *New York Stock Exchange Composite Transactions*, Wall St. J., Oct. 9, 1989, at C4.

of their prejudices. Apparently, the relationship between R&D expenditure and stock price valuation is much more sophisticated than a simple "increase your R&D and the market will knock your stock price down" correlation. Indeed, I suspect that the easiest way for Merck to slash its stock price would be for it to cut back dramatically on its R&D.

Further support for the view that the stock market does not invariably penalize increased R&D expenditures is found in a recent study by Randall Woolridge, who examined the stock price effects of announced changes in R&D budgets.[59] Woolridge found that in the two days following announcement of increased R&D budgets by 45 companies the value of those companies' shares increased by an average of 1.2 percent, net of overall market changes.[60]

For example, after du Pont announced on August 12, 1983, that it would spend an additional $100 million on R&D to improve automotive and industrial coatings, its stock price rose 2.54 percent. The market did not penalize du Pont with a decline. Similarly, a study by SEC economists found that a sample of 62 R&D announcements were associated with significantly positive stock price returns.[61] A study of the stock price effect of 658 announcements of changes in planned corporate capital expenditures also found that announcements of increased capital expenditures are correlated with significantly positive stock price effects while reductions in capital expenditures are correlated with declines.[62]

No doubt, these average statistics mask significant mistakes on both sides of the R&D fence. For example, when Federal Express in 1984 announced its plan to spend $1.2 billion over ten years to develop its Zapmail service, the stock market was as enthusiastic as Federal Express' management and the company's shares rose 2.27 percent.[63] Time proved that Federal Express and the stock market were both wrong about the promise of Zapmail—but the market's initial response was hardly hostile to management's long-term and expensive technology gamble.

Similarly, Genentech was able to raise $40 million in its initial public

---

[59] J.R. Woolridge, *Competitive Decline: Is a Myopic Stock Market to Blame?*, 1 J. Applied Corp. Fin. 26 (1988).

[60] *Id.* at 31.

[61] G. Jarrell, K. Lehn, and W. Marr, *Institutional Ownership, Tender Offers, and Long Term Investments*, Office of the Chief Economist, Securities and Exchange Commission (April 19, 1985).

[62] J. McConnell and C. Muscarella, *Corporate Capital Expenditure Decisions and the Market Value of the Firm*, 14 J. Fin. Econ. 399 (1985). The sample in this study had only eight announcements of changes in R&D expenditures and did not generate statistically significant results for this subsample.

[63] Woolridge, *supra* n. 59, at 33.

offering at a time when it had no meaningful revenues—much less profits.[64] All Genentech had was the dream that one day it might be able to develop useful products that might gain FDA approval and that might earn a profit for investors. The dream was, however, a distant one viewed from Genentech's initial public offering and, as events have subsequently proved, the gamble has not worked out as well as many scientists and investors had hoped.[65]

Thus, there are several examples of situations in which the market has been willing to reward high-intensity R&D companies with rich stock-price multiples and start-up funding. However, these are not the situations in which friction is likely to arise. Friction arises when management wants to pursue an R&D project or capital expenditure plan that the stock market won't support. In that situation managers often scream that the stock market is crazy and that it is only because of the market that they can't engage in valuable new investment.

But when management and the market disagree is it invariably true that the market is crazy and management wise? I think not, and there are several examples of projects in which management was willing foolishly to spend hundreds of millions of dollars despite the market's clear warnings to the contrary. My personal favorite example of a management that wouldn't listen to or learn from the market is the management of Unocal and its devotion to a shale oil conversion project that could be feasible only at sky-high oil prices. What Unocal's management overlooked, however, was that before the price of oil could rise to $50 or $60 a barrel, there would be so many other alternate sources of energy and conservation called on-line that the demand for high-priced shale oil might be problematic even if the technology was feasible. Neither the market nor I believed that this project made any sense. Nonetheless, Unocal's management pumped enough money into this project—both taxpayer dollars and shareholder dollars—that its expenditures substantially depressed its stock price and became a major magnet for Boone Pickens's attempted takeover of the company.[66]

---

[64]See Investment Dealers Digest, Oct. 21, 1980, at 10.

[65]See, e.g., C. Bartlett, Jr., *Special Situations*, Forbes, June 26, 1989, at 266; R. Stern and P. Bornstein, *Why New Issues Are Lousy Investments*, Forbes, Oct. 2, 1985, at 152, 154; *Waiting for a Payoff in Biotech Stocks*, Fortune, Nov. 26, 1984, at 185, 186.

[66]See, e.g., *Coming Up Dry: Unocal Struggles On With Attempt to Get Crude Oil from Shale*, Wall St. J., May 14, 1989, at 1; *Synfuels Corp. Issues New Grant, Defies DOE; $550 Million Approved for Oil Shale Profits*, Wash. Post, Oct. 17, 1985, at A21; *Lawsuit Is Filed to Void Accords for Unocal Plant*, Wall St. J., June 6, 1986, at A9. In addition, McConnell & Muscarella, *supra*, n. 62, observe that in the late 1970s announcements of increased expenditures on oil and gas exploration were correlated with stock price declines. The market was apparently signalling that exploration expenditures were not profitable given the outlook for petroleum prices and the availablity of

No doubt, there have also been and will continue to be situations in which managements would have supported successful R&D projects but for the market's skepticism. Similarly, there are probably R&D projects the market would have supported, but for management's lack of vision or courage in proposing them.

However, my goal is not to prove that markets are always right and that managements are always wrong. Instead, my goal is simply to get across the message that markets are not always wrong and managements are not always right. In particular, managements could often do themselves and their companies a great service if they just took some time to appreciate why the markets value some forms of long-term investment and penalize others. That simple step of market appreciation could probably work wonders for R&D budgeting, capital budgeting, and several other critical corporate decisions.[67]

## CONCLUSION

In sum, America needs to do more, much more, to strengthen and preserve its critical position in R&D. Stifling restructuring is not, however, the answer to America's R&D problem. Even if we brought restructuring to a screeching halt, our capital costs would remain far above our international competitors'. Moreover, most of the restructuring we would prevent would involve companies that do little or no R&D, and many of the companies that would be restructured are R&D laggards who spend less on innovation than their industry peers.

What then are we to do? A two-step program appears necessary.

First, we should take strong and immediate steps to reduce the cost of capital for R&D projects. Most fundamentally, America needs to increase its savings rate so that more domestic capital is available for R&D and for other investment projects. On a more targeted basis, R&D tax credits and reduced capital gains tax rates can also help lower the effective cost of capital for R&D projects.

Second, we must focus more of our efforts on commercialization. Japan is eating our lunch not only because its capital costs are lower, but because it has mastered the art of commercialization. All too often the United States stands at the cutting edge of R&D only to watch Japanese and other foreign firms earn the lion's share of the profits. This is not

---

reserves that could be purchased at lower cost. This divergence between market valuations and management expenditure plans can help explain much of the takeover activity in the oil and gas industry in the early 1980s. By the way, in this situation it appears that the market's valuation did a better job than management's plans.

[67] *See generally*, A. Rappaport, *Creating Shareholder Value* (1987).

smart R&D. This is charity R&D that does a disservice to the corporation, to its stockholders, and to its scientists who won't be able to do future R&D for that company unless it starts earning some profits from its past R&D efforts.

Neither of these two steps is easy. Neither is dramatic. Neither will satisfy the critics of restructuring who want to stop change with any argument they can find.

Either of these two steps will, however, help restore America's competitive R&D edge. If that's what we really care about, then that's clearly the direction in which we should go.

DR. PRESS: This symposium is designed to present both sides of the issue. For those of you who think that the keynote speaker is overwhelmingly biased toward one side of this issue, let me assure you that tomorrow you will hear the other side from our luncheon keynote speaker and during the course of the afternoon.

It is rather late to take questions and since we have an early start tomorrow, may I ask you to reserve your questions for the discussion period tomorrow. And since this was such an erudite talk, full of statistics, I think you need a little bit of time, those of you who want to ask your questions, to prepare adequately.

# Morning Session
## October 12, 1989

DR. WHITE: Good morning, everybody. I am Bob White. I am the President of the National Academy of Engineering, and it falls to me this morning to welcome you here to our symposium on corporate restructuring and industrial research and development.

The symposium addresses an issue that has been of nagging concern in the scientific and technical community. There is undefined anxiety that the recent turmoil in financial markets is or may be disturbing many corporations' ability to pursue traditional long-term research. The Academy of Sciences, the Academy of Engineering and the Institute of Medicine worry a great deal about the health of the R&D enterprise in this country, whether in government or universities or in industry and, of course, that is the reason why through the Academy Industry Program, we have decided to organize this program. We approach the issue with an open mind.

We do not believe the case is either proven or disproven. Our purpose here today is to hear differing views on the issue. Last night, we heard from Joseph Grundfest of the SEC that in his opinion, restructuring via takeovers, mergers, buyouts, is really not affecting the amount of R&D in the nation. He believes the real problems for the R&D enterprise, as you heard last evening, are perhaps elsewhere: cost of capital, things of that nature, which affect the investment in research and development in industry.

That is one view, but the state of play that we do face and that has given rise to the concerns that we have is that the past decade has witnessed intense restructurings, with record annual numbers of bankruptcies in the early 1980s, feverish activity in starting up new companies—for example, 200 biotech companies established between 1978 and 1982; peak numbers of mergers and acquisitions; in 1985 and 1986, the emergence of the leveraged buyout; the growth in foreign ownership of U.S. companies. Some states have begun to adopt legislation restricting takeover activities and the Congress is also considering issues and policies in this area and, in particular, changes in the tax code.

Today's symposium will, I am certain, refer to many of these developments, including some of the legislative proposals that are now being considered. As you know, the impact of corporate restructuring on R&D is an open question. Ken Flamm pointed out in his background paper for this symposium that we really do not have enough definitive data to know what determines the level of industrial R&D spending and economists who have studied this issue disagree. The symposium today is intended to bring together three separate but interdependent communities, whose collective decisions combine to determine much about the state of U.S. industry; Wall Street, corporate headquarters and public policymakers.

An effort has been made to have balanced views on each panel. The moderator for today's symposium has had much experience analyzing trends and sorting out policy options. As most of you know, Stuart Eizenstat headed President Carter's domestic policy staff in the White House. He is now a partner at Powell, Goldstein, Frazier and Murphy and he is also an adjunct professor and lecturer at the Kennedy School of Government at Harvard. He holds degrees from Harvard Law School and the University of North Carolina and is the author of frequent articles in many of our prominent newspapers and magazines.

On behalf of the National Research Council, I would like to thank him for agreeing to moderate today's symposium.

Stu Eizenstat.

MR. EIZENSTAT: Thank you very much, Bob, and welcome to all of you. With you, Frank Press, my friend Roger Altman and me here today, it proves that there is life after the Carter Administration.

I am also pleased that we have members of the Bush Administration who will be joining us. Jay French Hill, Deputy Assistant Secretary for Corporate Finance at Treasury; Rich Endres, Deputy Assistant Secretary for Technology Policy at Commerce and Bob Gray, the Associate Director of the Office of Management and Budget (OMB).

In the audience, too, during the day will be staff from Capitol Hill, as we had last night, as well as members of the press. But the most important part of the audience is its largest component and that is the men and women engaged in corporate R&D. When all of our talk about public policy options, strategies and global economics is over today, you will still be the ones who hold the key to the future competitiveness of U.S. industry.

The promotion of American research and development is an obviously crucial ingredient in our ability as a country to compete in a fiercely competitive global market. Fully 70 percent of our products must compete abroad or domestically against imports. We no longer have a monopoly on quality in this country and we can only stay ahead by being at the cutting edge of change through increased research and development.

There is a chronic systematic underinvestment in R&D by American industry. In part this is because of the recent uncertainties involved: For example, it has been estimated that 80 percent of all R&D projects fail to produce an economic profit. In part it is because firms can't fully capture the rewards of their own successful innovations: It is estimated that the rate of return to the public is twice that which accrues to the company actually doing the research, even when it produces an economic profit. As Joe Grundfest reminded us in his keynote address last night, if one takes non-defense corporate spending on R&D in our country, we are spending only about 1.9 percent of our GNP, compared with Japan at 2.7 percent and West Germany at 2.6 percent.

The general prosperity that the United States has enjoyed since the end of World War II is clearly based on many factors, but chief among them has been our ability to develop, create, manufacture and sell new products that to this day dazzle the imagination. Two of them come to mind, chosen, I must say, not quite at random: instant photography, Polaroid; and Tagamet, SmithKline.

These products and others like them have quite literally revolutionized American life in the postwar world and have been the sustaining force that has propelled our economy. I recently completed a book, Dan Boorstein's *The Democratic Experience*, in which he describes very graphically the way in which our innovation and inventiveness have created a more egalitarian democratic society. And, indeed, experimentation and innovation have been at the heart of domestic American history from the start.

Products like Polaroid's and SmithKline's and others like them have quite literally, therefore, been the sustaining force that has propelled our economy, but as everyone in this audience knows, we are in a new and much more challenging era. The world economy has changed dramatically since the halcyon days after World War II in which our country dominated the world economy.

The signs are evident all around us. Our share of world exports has continued to decline as first Europe and Japan and the newly-developed industrialized countries develop products in an export orientation. We have become so fixated on our own huge domestic market, that we have failed to produce products for a world market in which other nations have shown an increasing ability to match our quality at lower costs.

Our rate of productivity growth has declined each successive decade after the war. Our unit labor costs are among the highest in the world and we find that U.S. companies are moving production and R&D facilities abroad and sourcing out an ever-increasing percentage of components, even in so-called American products. Two of the four largest exporters, for example, to this country from Taiwan are not Taiwanese companies, but rather U.S. corporations, and the largest exporter of computers from Japan to the U.S. is an American, not a Japanese company.

The value of imported auto parts bought by our big three automakers more than doubled between 1982 and 1986, accounting for more than 5 percent of our 1986 trade deficit. Our triple digit budget deficits and the current accounts and trade deficits, which those deficits in turn have spawned, are, in a way, testimony to our own unwillingness to pay for our societal demands, to match our resources to what we require from government. This will affect our foreign policy and our influence abroad— for example, last year for the first time, Japan spent more money on foreign aid than we did. And it will also erode our capacity as a country to invest in education, training and health care, so important to a productive work force.

As a consequence of these changes, since the oil shock of 1973, real average household incomes have not increased at all over a 16-year period, even though we have far more two-earner families. For the first time, the younger generation's standard of living, when they become adults, may not be better than that of the older generation.

Hourly pay for production workers in this country was twice that of Japan as recently as 1985 and far above that of any other European country. Now, in U.S. dollars, it is lower than Japan's, West Germany's, France's and even Italy's.

Only through accelerated industrial research and development can we remain competitive and be at the cutting edge of change, staying ahead in the development of new products and production methods. The more R&D we can encourage, the more new and innovative products we can develop and the more productivity we can achieve.

We may disagree, as we will see today, about the causes of our concern and, indeed, about the solutions to what we have come to call America's competitiveness problem, but there is a shared belief that we have to do something about the future course and direction of high technology in this country. Indeed, even in this crucial high-tech industrial sector, so central to determining our standard of living, we now run a trade deficit with the rest of the world. Our semiconductor industry is on its way to dependency for its basic infrastructure on foreign suppliers, from silicon to etching equipment and quartz plates to wafer steppers.

A lot is at stake in the topic chosen for this conference, as we will be exploring the impact of corporate restructuring on industrial R&D. Does it further deter American research and development or is it a spur to it or, indeed, is its impact basically neutral? We know less about the answers to these questions than we should and it is my hope that we will generate enough heat as a result of your questions, the points raised by our speakers and future studies this symposium will spawn, that we will in the future be able to shed more light on this issue.

For the purposes of today's symposium, we should define restructuring quite broadly. It is, of course, at the very least, leveraged buyouts, but it also refers to mergers and acquisitions, takeovers, both hostile and friendly, as well as internal restructurings, which most large corporations have gone through in recent years.

Just to cite five R&D-based American companies, all members or former members of the Academy Industry Program, which is our sponsor today, which have experienced dramatic, if different, restructurings during the past five years: AT&T, which was effectively broken up; General Electric, which acquired RCA; Eastman Kodak, which acquired Sterling Drug; Hughes Aircraft, acquired by GM; and Dupont, which acquired Conoco. Any daily edition of a major newspaper dramatizes how tumultuous the corporate landscape has become.

Just a random example in the September 29th *New York Times* had headlines proclaiming, "Dupont, Merck, Set Drug Pact; Kyocera and AVX Agree to Merge in a Stock Swap; Borden to Cut 7,000 Jobs in 65 Plants."

The competitive pressures of our new global economy are affecting every major corporation in this country, not only the ones, in Wall Street's terminology, that have been put in play, bought out or acquired. All are restructuring to one degree or another or they know they can't survive. The question then remains: What is the impact of all this activity on long-term growth? Has this concern for performance in the next quarter caused myopic managers to cut expensive and risky research, which may not pay off for years to come? Or alternatively, does our new competitive environment lead to better management and more intelligent investment decisions? These are among the questions that we will be exploring today.

Aggregate R&D in this country is about $131 billion, half of which comes from the federal government, a figure including military research. I think we have all read the President's lips sufficiently to know that we are not likely to have a great plethora of federal money in the intense competition for federal dollars. While NSF budgets are going up modestly, we can't expect explosive growth in this fiscal environment in research dollars coming out of Washington. So, the question is: Will the private sector make up the slack?

As *Business Week* editorialized, with the U.S. under escalating pressure from competitors, an R&D slowdown could not come at a worse time. A poll of this country's 476 largest industrial corporations revealed that most of the firms' R&D budgets would not be growing as fast in the next three years as they had in the previous three. And the latest Industrial Research Institute survey of 161 R&D directors found that R&D funding as a percentage of sales was expected to continue to drop.

This symposium will be divided into four parts. First, we will hear two views from Wall Street, from Roger Altman of The Blackstone Group and Michael Tokarz of KKR, who will tell us how money managers, who are very often the catalysts for change in our economy, view the current situation.

We will then turn to two senior corporate executives: Mac Booth, President and Chief Executive Officer of Polaroid and Henry Wendt, Chairman of SmithKline, who will describe their respective companies' restructuring experiences and their likely impact on R&D.

After lunch, the Majority Leader of the House of Representatives, Dick Gephardt, will outline how Congress might try to deal with this changing corporate landscape.

And finally, three economists will outline their views on what is going on in general.

Before introducing the first panel, I would like to take this opportunity to thank the Academy Industry Program staff for their work in putting this program together; in particular, Ed Abrahams, who has been extraordinarily creative, helpful and indefatigable in helping to put these panels together and working with me. I would like to personally thank Frank Press and Bob White, the Presidents of the National Academies of Science and Engineering, for encouraging a full balanced examination of this critically important topic.

With no further ado, I would like to start by introducing a long-time friend and former colleague, who will be our first speaker, Roger Altman. Roger is Vice Chairman of The Blackstone Group, a private merchant banking firm, which advises major corporations on financial and corporate development matters and also employs a large pool of capital for buyouts and other principal investments.

Prior to joining Blackstone, Roger was a managing director of Shearson Lehman and from 1977 to 1981, worked with us as Assistant Secretary of the Treasury for Domestic Finance with responsibility for federal, municipal and corporate finance. While at Treasury, among the many issues Roger dealt with were the Department's own borrowing program, a particular challenge during our era; the Chrysler bailout; and the rescue of New York City. It seems like ancient history, but, Roger, you were very much a part of making history, which is, in fact, very topical.

Roger was educated at Georgetown University and the University of Chicago, from which he received a master's in business administration.

Roger, if you will start, I will introduce Michael afterward, and thanks so much for coming, all of you for coming.

MR. ALTMAN: Thank you, Stu, and good morning, everyone.

Let me say when Stu originally asked me if I would address this Academy Industry Program of the National Research Council, I first thought that he had dialed the wrong number, that he intended to call someone with an affinity for science and somehow mistakenly got me, because if you knew my true ignorance on science, engineering and medicine, you would appreciate my own confusion about the invitation.

But when Stu assured me that it was not his speed dial run amuck, I set about thinking what I could possibly contribute this morning. Now, I still don't have a completely clear answer to that, but I hope that when I am finished, it will be a positive rather than a negative contribution.

I also should say at the beginning that I am perfectly willing to take a polygraph test to certify that I did not attend last evening's event and did not hear Joe Grundfest's comments. And I say that because I just looked over his speech this morning and you will see in a minute why I assure you that I didn't have a chance to see his comments in advance of my own.

Let me start off then by suggesting that we look at this question of the impact of corporate restructuring on industrial research in a little bit broader context. I am sure most of you or all of you have read Mr. Flamm's excellent paper, which concludes, in my view, correctly, that there is no conclusive evidence of a negative impact of restructurings on industrial research.

Now, while many of us believe intuitively that restructurings and their attendant leverage are bad for sustaining, indeed, for expanding, as Stu has correctly observed that we must, necessary levels of R&D, the direct evidence is not there. There are a number of logical reasons why that is the case and I will address those toward the conclusion of my comments, but I would like to start by suggesting, as I said, a broader context for the question itself.

It seems to me that restructurings are one part, and they are not the main part, of a larger financial trend in the United States, an economic trend, and a very negative one, which is the savings and investment crisis, the dearth of savings and investment, and its impact on the cost of capital. We have seen over the 1980s such sharp declines in private savings and in net private investments and such a widening divergence in the cost of capital for American industry, compared with that of our key trading partners, that restructurings, in my view, have been an inevitable consequence of those trends.

But they are not a cause of it. The reason they are a consequence is that as the cost of capital differentials have widened—and you have seen Joe Grundfest's rather startling statistics from the New York Fed as to how wide they really are—obviously, only investments with faster paybacks and higher returns can be rationalized. That has given rise to this near frenzy in corporate restructurings, which, as Stu correctly points out, really encompass a rather wide variety of corporate situations, extending far beyond leveraged buyouts.

So, I would like, at the beginning, to just spend a few minutes, and I will try to avoid repeating too closely what Commissioner Grundfest said, looking at this larger context. It seems to me that whatever you conclude today on this question of restructurings and industrial research, we are not going to be in a position to expand the contribution of research and development to gross national product (GNP) in this country, or to any other measure of R&D's contribution, without a very fundamental change in both public policies and private behavior, as they relate to savings and investment.

Now, you probably all know about various trends in terms of savings and investment, but maybe you don't know quite how serious or how bad these trends have been. Let me say that I had an interesting experience a few months ago, which put this in vivid context, at least for me. I read an editorial in *The Wall Street Journal*, which, obviously, is the most widely read business publication in the world, to the effect that we were experiencing a savings and investment boom in this country.

For those of you who read *The Wall Street Journal* editorial page, and most people I know don't admit it if they do, you know that has been their long-standing view for the last couple of years. I have a friend who is on the editorial staff and I called him and I said, you know, I really think that that is wrong, that, in fact, we are experiencing sharp declines in savings and investment.

And after several weeks of debating—and I am not in the business of research myself, even financial research—I managed to get my hands on a variety of Federal Reserve of New York-type analyses. They agreed, at least, to run an article which I wrote, to the effect that the truth was really the opposite. My point is not that I am right and they are wrong. My point is that there is a widespread view in many sectors of America, including very influential sectors, that there is no problem.

Let me illustrate why there is not only a problem, but a grave and a worsening problem. For the 30 years ending 1980, private savings in the U.S. averaged about 7.2 percent of all income. This decade, through 1988, the figure has been about 2.8 percent. Net private investment—now, there are a lot of different ways of measuring net private investment, as *The Wall Street Journal* almost endlessly pointed out to me. But net private investment, excluding housing, according to the Commerce Department figures, averaged 3 1/2 percent of GNP during this earlier 30-year period and it was quite steady in the fifties, the sixties, the seventies.

This decade, it has averaged approximately 2.2 percent, which, by the way, is the lowest level since the Second World War. So, it should be no surprise to any of us that productivity growth has fallen below 1 percent, or that, as Stu alluded, real per capita income for the bulk of Americans is stagnant and for a very high percentage of wage earners actually declining, or that long-term investment, of course, just like industrial research, is weaker.

Now, to put that into today's context on restructurings, let me turn to this question of cost of capital and how the cost of capital disparities spring really quite clearly from these savings and investment declines.

Three years ago, I had the opportunity to teach a course at Yale for a year, on issues that included very much the question of finance and finance comparisons, comparing the financial systems of, and the related impacts on, the U.S., Germany, and Japan. We had a chance to look at this in great depth. The disparities since that time have only widened.

One way to look at it, which Joe Grundfest did not mention, is to examine returns on investment and returns on equity. If you use the U.S.-Japanese comparison—and it is the most vivid and, of course, the most relevant in a lot of ways—you see that industrial returns on investment here are considerably higher, have been considerably higher for most of the 1980s. I have seen some studies, including a very good one that Professor Ellsworth at Harvard did a couple of years ago, that equity returns in a variety of major U.S. industries have been more than 50 percent higher than the Japanese counterparts, and particularly using industries where most of the relevant assets are in the home countries.

There are a few Japanese industries that are still extraordinarily leveraged and where equity returns are therefore somewhat distorted. But generally returns on equity are much, much higher in the United States than they are in Japan and, for that matter, in many other nations against whom we compete. Now, if you add to that equity point the fact that, as we all know, U.S. interest rates have been consistently higher during the eighties than those in, for instance, Japan and Germany—dramatically higher—you understand then that the cost of capital here is as much as 100 percent higher than in Japan and perhaps 50 percent higher than in Germany.

Now, any of you who are directly familiar with the discount rates, which Japanese acquirers use—or Japanese investors, I should say—in evaluating investments, you know what I am referring to. Let me just tell you a quick anecdote, which perhaps illustrates it well.

Two years ago, our firm was representing a major Swiss company in its efforts to acquire certain Canadian assets, which had been acquired a bit earlier by a Japanese buyer. The Japanese buyer wanted to keep most of the assets but was interested, we understood, in divesting some of them.

Well, we all know that the Swiss enjoy relatively low borrowing costs and, of course, are in a position to evaluate investments with that advantage.

In this particular case, we had made quite an aggressive offer. I think we had calculated the discounted cash flow valuation in this case, using about an 8 percent discount rate, which I think any of you in the corporate sector would say is a lot lower than you would use here in the United States.

After a few hours of negotiating we had gotten essentially nowhere, having been told time after time that our offer was way too low and that the business was worth much more than that. It was a sufficiently amiable context that I finally asked if they could explain to us how they got their valuation, because we just couldn't understand it. They were quite open, unlike some of the Japanese negotiations I have been in, and actually handed over their work papers and showed us how they made their calculations. There were these smudged worksheets and we hovered over them, kind of peering at them and, sure enough, right there was the discount rate they had used. It was 4 percent. So one of us asked how can you use a 4 percent discount rate, and they said very simple: That is our true cost of capital. Well, needless to say that meeting ended promptly thereafter.

The point is, of course, that if you are looking at the world through the perspective of capital costs in that range, then obviously you are able to undertake investments with longer breakeven periods or payback periods and fundamentally lower rates of return. Needless to say, it is a dramatic competitive disadvantage over the long term.

There are a lot of reasons why the cost of capital has risen so much in the United States. I won't dwell on them in great depth because that is not our focus today, but I would like to just tick them off as I see them.

One is the legacy of inflation and economic volatility of the seventies and early eighties, because it changed expectations on risk. As risk is perceived to rise, of course, returns must be higher to compensate.

Another reason, as I mentioned, is that the pool of private savings in America has dwindled during the eighties. Obviously, the competition for those remaining funds has driven up their cost and, of course, our persistent high budget deficits and the corollary current account deficits have been a key cause for that decline in private savings.

A third reason is that the U.S. equity markets today are really totally dominated by performance-driven institutional investors of all kinds. The individual investor, for all practical purposes, is a bystander. As real investment rates have stayed high, and with investment managers compensated in line with performance, the drive for higher equity returns is insatiable. One manifestation of that is the almost endless stream of exotic financial products that we see produced as a way to try to capture higher returns.

Maybe the greatest source of creativity in this country, or certainly one of them, is the financial sector, in terms of that exotica.

Last, the U.S. banking system has been profoundly changed over the last two decades. Larger U.S. corporations now bypass banks. They borrow directly from a variety of different public credit markets at home and abroad, from commercial paper all the way to long-term markets. Corporations no longer have close relationships with banks or other lenders and the term "committed lender" has almost become an oxymoron in the U.S. Contrast that to the universal banking systems in Europe and Japan, where, among other things, banks are permitted to be major shareholders in corporations, which, of course, breeds long-term lending relationships, the type of thick or thin relationships that you used to see in the United States 25 years ago. But such shareholding relationships, such interlocking ownerships, are prohibited in the U.S. And so, for that and other reasons, we have evolved a very different banking system in this country than our major competitors have.

And I think we have gone from the point where it was a matter of pride—and correct, too—to say that the U.S. capital markets were the broadest and deepest in the world and a major competitive advantage for the United States in industrial development, to the point where the opposite is relatively true. The U.S. financial system today is not a competitive advantage for this country as it relates to industrial development.

Now, as to the implications of this world in which savings are low, investment is low, cost of capital is high. I think some of them are directly relevant to the question that is the direct topic for today—restructurings. I mentioned that investment horizons have shortened and shortened in America. That applies to capital budgeting decisions, as I said; it applies to venture capital decisions; it applies to institutional shareholders of public stocks; and it applies to restructuring decisions taken by corporate managements and boards of directors.

Let me say a few words then on restructurings, why they are happening, what they mean and what the future will bring.

As I said at the outset, the term "restructuring" really should be applied quite broadly. It applies essentially to corporate transactions aimed at delivering higher values to shareholders. Now, there are also restructurings which relate to financial difficulty. Some people call them financial restructurings, of course. Some people call them reorganizations. They are, to some extent, in a different category because they relate to financial difficulties, but, if you will, you can include them in this topic as well.

But the main types of restructurings I am referring to have boomed in the last few years because, of course, shareholders are no longer patient as they search for higher and higher equity returns over major value gaps: the differential between private market value or what Wall Street calls M&A

value, and public market value. Now, those value gaps have become in the last few years, during the eighties particularly, quite wide in many cases. Time after time, major company shares have traded as low as half of private market value. Faced with that, outsiders have pressured such companies to close those gaps and deliver much of that value, the hidden value, if you will, to shareholders.

Basically, those gaps exist because investors are prepared to pay high premiums to truly control the cash flow. Instead of the dividend income, which an owner of a hundred shares looks forward to—either actually paid or anticipated eventual dividend income, and, of course, the capital gains potential—the owner of all shares controls the free cash flow and can apply it in any way he chooses.

Those value gaps also exist, as Professor Michael Jensen at Harvard has effectively pointed out, because managements often are motivated to manage the assets of the corporation in a way that is inconsistent with maximizing the present value of the cash flow or the present value of the shareholders' investment.

Another important development in this context has been the more and more sophisticated use of leverage. I think this is a particularly misunderstood part of the corporate restructuring trend. Let me try to point it out in a somewhat different way.

A couple of years ago, Goldman Sachs simulated the investment returns for the first seven years of the eighties, which would have been achieved had an investor selected a random portfolio of New York Stock Exchange stocks and then applied to that portfolio the type of leverage that is today achievable in private market transactions. For simplicity purposes, you can say that is about 10 to 1, debt to equity. The result, because of the sharply declining interest rates that we had from 1982 forward and because of the rising stock market over that same period, would have been about 70 percent average annual return on equity.

The point is that leverage, together with the rising stock market, has permitted leveraged buyouts and many other types of corporate restructurings to work as well as they have. And it has been in some respects so easy, that no one really should be surprised at the billions which have flowed into LBO pools—including, for that matter, our own firms—and at the interest that these transactions have attracted from investors.

I think a central point to your topic today is—and I will get to this in more detail in one minute—it isn't necessary in so many restructurings, I would argue in most of them, to squeeze every last dime of cash flow out of a corporation as if you were liquidating it. In most cases—and I think the results that Kohlberg, Kravis has achieved illustrate this—in most cases, the practice that investors apply is really quite the opposite. It relates to growing the core assets and capitalizing on that magic of leverage, rather than stripping the core assets or, as I say, squeezing them as if trying to get blood from a stone, to get the last dollar over the first year.

As for the future of restructurings, the restructuring movement continues to grow. I only see one obstacle that might present itself to that movement, and that is a possible recession. The last six years of the restructuring boom, of course, have not seen a recession. There are many arguments as to the impact of a recession on restructurings; most expectations are that it would dampen the restructuring movement quite substantially, but until that occurs, the corporations will increasingly adopt methods of unlocking values and delivering them to shareholders. Many corporations, in my view, are increasingly going to do that on a preemptive, proactive basis of their own volition, rather than being forced to by outsiders.

There will be more cases, for instance, like Quantum Chemical, a major producer of basic chemicals—it used to be called National Distillers—which several months ago voluntarily chose to recapitalize itself, pay a special cash dividend to its shareholders, to close the gap between its private market value or its leveraged value, and its then desultory stock price. I think you will see a lot more of that.

Finally, let me come back to the beginning. Mr. Flamm's paper concludes that there is no evidence that the research intensity of companies which have been restructured is lower than that of other comparable companies which have not been restructured. I think that is logical for three reasons.

First, truly high-technology companies rarely get leveraged up through restructurings. In so many cases, technology is too volatile and the predictability of operating performance is too low. Putting it very crudely, they don't make good LBOs.

Second, investors in restructuring are primarily looking to what Wall Street calls the exit; in other words, that they will liquify their investments over, say, a three- to five-year period by selling or refinancing the core assets, the central assets. That requires, of course, those assets be performing well at that time. So these investors do not have an incentive to cut R&D. It might increase cash flow a bit over the very short term, but it will hurt the exit price, which is what really counts in terms of achieving the returns that those investors seek.

As I said, most restructurings work through the economics of leverage, not by squeezing the cash flow in a liquidation style. Remember, if you buy an asset for 100 dollars and you borrow 90 and you put up 10 of your own and two years later you sell it for 120, you have doubled your investment. The value of the total business may only have increased 20 percent, but your equity has doubled. So, cutting back R&D is not usually a consequence of restructurings unless we are talking about a company which is really facing a reorganization.

Let me conclude then by making a prediction. It may not be a happy one, and unlike my World Series bets, I hope I am wrong on it, but I don't think so. The dearth of private savings and of net investment is going to continue in this country and the unusually high cost of capital for industry is going to continue and the restructuring trend, which it has caused, is going to continue.

Both public policy and private consumption trends and other types of private behavior in America, are harmful today to investment and they are harmful, it seems to me, to R&D. Without adopting a prosavings and a proinvestment set of policies, particularly tax and budget policies, this negative climate is destined to continue. Perhaps there is some consolation in the extent to which other industrialized nations also are seeing slowdowns, as they are, in their rates of capital formation, but ours, unfortunately, is much more pronounced.

Now, the hopeful side of it is, if I can say so, Stu, that with the talent assembled here and your obvious concern over these trends and that of others like you, perhaps these attitudes can begin to change and the pressure on R&D spending can be alleviated. All of us must hope so.

Thank you very much.

MR. EIZENSTAT: Roger, thank you for getting us off to a very good and provocative start.

Our next speaker will be Michael Tokarz, who is an associate at KKR & Company, which I think everyone would agree has truly revolutionized corporate finance in America during this decade. Since 1976 KKR has grown from a company with just $3 million of its own to invest, to a company that controls an empire of 35 companies at a cost of more than $65 billion. Together, these companies would make up the largest industrial conglomerate in America. They include not only giants like RJR Nabisco, Stop & Shop and Safeway, but also R&D-based corporations, like Duracell.

Michael joined KKR in 1985 and since then has participated in numerous leveraged buyouts, financings, restructurings and dispositions. He currently serves on the board of directors of Beatrice, Walter Industries, Safeway, IDEX and K3 Holdings.

Prior to joining KKR, he was a vice president at Continental Illinois Bank. He was educated at the University of Illinois, where he studied both economics and business administration. We look forward to hearing from you and appreciate you coming, Michael. Thanks.

MR. TOKARZ: Good morning.

In anticipation of being here, I was thinking of your perceptions of what we do. Generally, leveraged buyouts and how you might feel about them pervaded my thoughts. I decided it wasn't really fashionable for any of you to say "I think leveraged buyouts are great." After all, who do you know who is highly leveraged that is really in great shape. Even the word "leverage" has an awkward, negative connotation. Certainly, relative to words you are familiar with, such as research and development, which connotate building and growth, the words "leveraged buyout" stand in a stark negative contrast; a building process versus a leveraged buyout, which connotes a consumptive or a consuming process. Yet, I think, strange as it may seem, on a macroeconomic basis, leveraged buyouts and restructurings may actually be an agent for increased research and development. And why? What is happening?

Why is our productivity decreasing? Why are our labor costs higher? What is happening? Where is America in this competitive front with the Japanese and other competitive nations? Why are we behind? Why are our television sets made elsewhere? What is happening?

Well, it is easy to blame the leveraged buyout because we all were raised with the notion that leverage is bad. It is easy to blame the things we don't understand, as I am sure you can appreciate.

After all, if 25 percent of our college seniors don't even know in what century Columbus found America, imagine how their lack of understanding, and the general lack of understanding in our society, breeds fear of change, fear of the research you do, fear of the change we are in the process of stimulating ourselves.

Interaction between our communities—research and development and financial communities—is quite infrequent. Therefore, our understanding of each other is limited and quite possibly we fear each other.

The financial markets are in a rapid revolution, rapid change. Allocation of capital is coming under increasing demands for change, rationalization, redeployment. Today, you will be given or you can pick up on the way out an article written in this month's *Harvard Business Review* by Professor Jensen of Harvard, which deals with something that I think many of you are going to find hard to understand. That is, that the American corporation, the American financial system, the renowned engine of world and global growth, is actually fat and inefficient, poorly structured and badly in need of change.

I sit there and I say what is it that you do? What is it that I am afraid of, why I look at you and cast a wary eye on all of you research and development people, people in the sciences. I wonder about your gene splicing, your cloning, your bioengineering, and, most of all, I find it hard to be able to eat the strawberries you sprinkle that beneficial bacteria on.

What are we going to do together as a society? What part do I play? What part do restructurings play and what part do leveraged buyouts play?

Well, I am afraid I don't have the answers to that and as Joe Grundfest and Mr. Altman have briefly explained already through their comments, we are not sure what to do. No one really knows what is happening. But I can assure you of one thing, that what is happening in the finance industry is an agent of change that is producing results that I think you are going to find interesting. What are buyouts doing? Where have buyouts been done?

Well, they are typically done on stable, generally unleveraged companies that are not R&D intensive at all. The evidence produced by Joe Grundfest and Marty Altman, with which I won't bore you by going over it again, shows that we don't buy R&D intensive companies at all. As a matter of fact, we generally avoid them.

So, where are restructurings happening? Well they are happening in inefficient or large corporations that have limited opportunity to deploy their capital and their businesses in high-return or marginally beneficial projects. High-return projects are used up, in essence, in areas such as food, retailing, banking and tobacco.

As a result, these mature corporations have accumulated massive amounts of capital and, as Jensen points out in his study and Kaplan, from the University of Chicago, points out in another study, 40 percent of the production of American corporations between 1940 and 1980 was wasted. Forty percent of the productive capacity, the assets generated that could be redeployed into R&D or elsewhere, were wasted.

They were wasted because managements were paid on growth, size and the volume of assets they had under their control. Companies husbanded capital rather than disseminating it to shareholders and the financial markets, limiting redeployment into projects with higher returns.

How is it that a leveraged buyout or a management restructuring opportunity can pay on average a 50 percent premium over the public trading value of a company? How can we attract the capital? Who will give us the money to go and pay 50 percent more than what the so-called sophisticated and perfect knowledge markets are paying for a business?

Well, again, Roger pointed out that it is corporate control. It is the ability as an entrepreneur, as an owner, to take a set of assets and manage them more efficiently. In short, drive the race car faster and more efficiently than the next guy; same asset, different management, different result.

Cost of capital—macroeconomic statistics, more than we can possibly digest. But isn't it supply and demand? Isn't it true that capital is, in part, allocated through a system, and isn't the system today in America a function of the public corporation? The public corporation: RJR Nabisco, trading at $50.00 a share, public market value about $15 billion. Management decides that they can buy that company and productively manage it under private enterprise. So, they bid approximately $20 billion. The financial markets take over and we end up buying it for a total capitalization in excess of $30 billion.

Many of you look at your R&D budget and say just give me a little, just to have a chance to deploy that; we can show you how productive we can be. Well, in essence, in a macroeconomic sense, that is just what we did.

The capital locked up in a tobacco company and in a food company—tens of billions of dollars—was returned to shareholders, those people who had equity investments, those people who will take that money, pay their taxes and then, again, reinvest it, seeking out yet another new opportunity.

I can assure you from the work we have done in organizations such as Beatrice, Safeway, RJR, that the internal utilization of cash, prior to our purchase, the internal utilization of the capital was, as Jensen puts it, inefficient.

I would like to draw a quick parallel as to why. They key is management, if you take a look at the capital accumulation in our country, the great growth, the industrial capacities built up in the late 1800s and early 1900s, they were done by owner/manager entrepreneurs. The Rockefellers, the Carnegies and many of those names you read in the history books built up our country into the capital and industrial giant it was. But they passed— their generation passed and they passed control and ownership through financial markets to the multitudes, to the institutions, to the individual investors and they passed the management of their corporations to the professional manager. The professional manager—the professional manager of the corporate asset, the professional manager of the institutional dollar, the savings dollar that we all talk about.

Well, what we have come to realize is a very simple tenet—and this is the predicate upon which we build our purchase and our belief that we can pay more for a company, 50 percent on average more than someone else is willing to pay, and do better with it. Again, I would like you to read Professor Jensen's article because it focuses on it: in our buyouts, ownership of the corporation is rendered back in part—a substantial part— to the management. They own it. They run it for the long term. Quarter-to-quarter earnings do not pervade the mentality.

Careers, month-to-month, quarter-to-quarter, year-to-year performance bonuses are not predicated upon short-term results. They don't ask: did I do better than last month, did I do better than the last quarter? They don't think: Oh, my God, if I make this long-term investment in an R&D expense, which is a current expense, charged currently to the quarter, if I make that investment, my earnings go down; I look bad. I won't do as well and my career may be impaired because the return may be five, seven, ten years off.

Yet, when we do a leveraged buyout, management is building value for the long term. I can assure you Duracell's R&D budget is dramatically higher under private ownership than it was under public ownership. Our capitalization is much more leveraged than Kraft, the prior owner of Duracell. But why?

Because management runs the company and we are the owner and we want long-term value appreciation. In the process of reallocation of capital to change the public corporation, the agent of change is the people—the Boone Pickens, the Sir James Goldsmiths, and some of these nasty characters—who attack these corporations that waste 40 percent of their assets. They are able to attract capital as we are—$5.6 billion in our case—to invest in these companies because of the simple difference between a renter and an owner. I characterize management of many companies today as mere renters of those corporate assets; that psychological difference between ownership and renting, we think, drives the long-term considerations in many of our companies.

Rather than cover many of the macroeconomic statistics that were covered by the speakers you have already heard from, I am going to alter my talk. So you can have a little understanding of what we do, I am going to talk a little bit about buyouts, how they are structured and some of the results we have obtained in examining our own records in areas such as research and development, capital spending. I will spend a few minutes on that.

The definition of a leveraged buyout, similar to restructuring, is that a leveraged buyout is nothing more than a financing technique. Its name is driven by the fact that the acquirer usually borrows a large percentage of the purchase price, typically from a variety of sources, such as commercial banks, insurance companies and other sophisticated financial institutions or public purchasers of a high-yield financial instrument often called junk debt.

Occasionally, a portion of the purchase price is paid in the form of securities, which are distributed to the public, who originally owned the company. Generally, in a leveraged buyout, a company goes from being a public company to one that is privately held by a small group of owner-investors; such are the investors in KKR Funds.

What is the capital structure of the buyout? Well, you hear about these wildly leveraged schemes. For most major corporations that we structure, we find that 40 percent of the debt for that company is in the form of bank loans; 25 percent in long-term subordinated, often called high-yield securities; 15 percent in the form of a preferred stock instrument, many times sold to investors; and then 10 percent in common equity that we may put in.

The critical element, again, is that ownership, the management of the corporation, puts up its own money and has a risk/reward profile approximately 200 times more effective than it would have if it were in a publicly-held company. For every 1,000 dollars of value increase in the company in a leveraged buyout, a manager has 200 times the return to him personally than he would if he were in a publicly-held company.

The criteria for our buyouts are fairly simple. The financial characteristics include such things as a demonstrated stable profitability history and the ability to obtain above-average profit margins over a longer period of time. Turned-around situations are not attractive. Strong, predictable cash flows to service the financing costs related to the acquisition are critical. Readily separable assets or businesses which could be available for sale to provide flexibility are often necessary.

The business characteristics: a strong management team. We don't provide management and as you will see in Jensen's article, most LBO firms or restructuring firms have very low component. At KKR, we have $63 billion of corporate purchases under our belt and only 16 professionals.

The companies we buy usually have well-known brand names and strong market positions. Typically, they are low-cost producers within industry and thereby competitive and competitively advantaged. We look for circumstances where there is potential for real growth over the long term and we hope our companies are not subject to prolonged cyclical swings in profitability or in their business segments.

And, most important for this group, products which are not subject to rapid technological changes. So, what do we buy? We buy a Safeway food store chain, $20 billion in revenues. Upon simple examination I am sure you would agree that if one-third of those revenues produced a loss then approximately one-third of those revenues are, in fact, inefficient. By dealing with the one-third of revenues that produced the loss, within two years we are able to double the profitability of the company. A short while thereafter, we are able to triple the investment in new stores, new concepts, in new delivery mechanisms.

In Beatrice, when we examined its budget, we found that a company called Tropicana—many of you may drink their orange juice—was denied the capital to complete a project and packaging, which would extend the shelf life, extend the quality and the Vitamin C retention of the juice for an additional four weeks. They were denied the capital to put into their organization a computerized system for manufacturing, a computerized system for corporate control and distribution of fresh juice from Florida to New York and the East Coast.

This was an A-rated company and its subsidiary was denied a meager $40 million to complete these projects. With a net worth of over $1.5 billion, with cash flow over $1.1 billion, they were denied the $40 million, not for one year, not for two years, but for three years in a row.

When we took over the company, we found that that $40 million had been allocated to a Mario Andretti race car program. As a matter of fact, it was a $75 million commitment over a three-year period to put the Beatrice name on the bonnet of a race car. The company had spent in its capital program over $80 million on corporate aircraft in the previous three years; it had spent $200 million developing a plan to advertise the name Beatrice and relabel all of its products so that the red label "Beatrice" was on the package.

When you look at the Mario Andretti program, you look at the Beatrice labeling program, the stunning fact was there wasn't a single product that was sold under the label "Beatrice," not one. Yet, $40 million was denied for the research and development of packaging and the computerized systems for manufacturing and distribution for one of its subsidiaries.

Well, if you look at the statistics for the capital spending of Beatrice, you will see that after we took over, it went down. Yes, we did not spend money on the Mario Andretti program. We cancelled it. Yes, we did cancel the Beatrice labeling program and we sold the planes, but we redeployed the money, $40 million of it anyway, into the packaging program, into the management information systems program for Tropicana.

It has been alleged that buyouts reduce tax—are based on tax benefits. I would like to cover that for a second. Do buyouts result in lost tax revenues for the government? We did a study and the University of Chicago did a study. On all of our acquisitions, prior to RJR, we have calculated that a net tax benefit to the government, present value, was $3.2 billion, positive. The Kaplan study, looking at the RJR transaction alone, shows a net present value to the government in tax revenues of $3.8 billion, for a total of that present value benefit to the government treasury of $7.0 billion as a result of leveraged buyouts that we did.

What is the effect of buyouts on employment? We have found that, and you will see in the Kaplan studies and in the Jensen study, that employment actually grows after a leveraged buyout because the companies are so much more efficient. In our own analysis, based upon budgeted numbers in part for the last year, we found that our annual rate of growth of employment went from 2.3 percent three years prior to the buyout, to 4.2 percent after.

We were criticized for this study, by the way—some of you may have seen *The Wall Street Journal*—because we included a budgeted number. So, we recast the numbers only for actual results, three years prior, three years after. The employment growth for the companies only three years prior was 3.4 percent; after, there was 8.0 percent growth in employment.

Can buyouts and restructuring stand negative events and downturns and recessions? It amazes me. I read the paper. I see all of the press saying, well, wait until these buyouts hit the next recession, boy, are you going to see a mass of problems. Well, we arranged 13 leveraged buyouts in the period from May 1, 1976 to December 31, 1982. All of these buyouts were owned through the recession of the early 1980s, which was one of the deepest recessions we have had since the Great Depression.

Prime interest rates soared to over 20 percent. Despite the recession, each and every one of these buyouts retired all of their debt on time or early and collectively provided the equity investors with an annual compounded rate of return of 35 percent. At the same time, I am unaware of any buyouts or leveraged restructurings in the television, steel, machine tool, auto, tire manufacturing, banking or electronics industries, all of which suffered dramatic declines in market shares and profitability during this period and many of which were notably bankruptcies or troubled situations.

What about capital spending? Well, I have told you already, we do cut out wasteful spending, but we have found out that in our companies, where management is in control and again an owner, that the rate of investment in capital projects increases over the years prior to the buyout.

What about research and development? In our companies, on average, research and development had a declining trend in the three years prior to our purchase. On average, after the period of purchase for the next three years, it had an increasing trend. Capital R&D spending was up 15 percent in the three-year-period postbuyout, relative to the period prior to the buyout.

The Kaplan study of LBOs showed, however—and this is important to know—that of all our companies and all the companies that do R&D that are leveraged or restructured, that only 1 percent of the revenues is really dedicated to R&D. Therefore, although the percentages may seem impressive, I have to submit to you, again, we do not buy R&D intensive companies. With $131 billion of R&D in the United States, I am afraid that KKR companies are responsible for only $250 million.

Is corporate leverage excessive in the economy today? Well, relative to our trading partners, we are the least leveraged. As a matter of fact, on a market value basis, debt to equity, we are at a 57 percent ratio today, the lowest ratio of debt to market value of equity in the last 20 years.

Finally, I would like to talk a little bit about whether companies are more efficient after a buyout or not. Well, we tend to make money, as you know, on average, but are they more efficient? Are we milking them? I think fundamentally all of you would agree with this example. If you bought a house and paid for it with leverage, and you wake up in the morning and realize that you own only 10 cents of every dollar in that house personally—the rest was borrowed—when you look out at the lawn, you are not going to neglect it. And if the roof begins to leak, you are not likely to let it go.

You are not going to sell the garden to the neighbor and you are not going to rent your garage to the kennel. Clearly, you are not going to take a bunch of strangers into your house and let them run rampant. You are going to take care of it because you know from the record that you can sell the home for more than you paid for it. Certainly, none of you would be able to sell your house, nor could I, if we let it go into a gross state of disrepair.

Jensen says that the pay/performance ratio of the ownership profile that I am talking about in these companies is very significant, very strong. The CEO of a public company owns less than 0.25 percent, less than a quarter of 1 percent of the economic value appreciation of his enterprise, versus a leveraged buyout manager, who owns 6.4 percent on average, a ratio of over 200 percent stronger. He cares about this house. That manager cares about how he spends on it.

As for performance over the long run, Kaplan of the University of Chicago did a study of over 96 buyouts in leveraged restructuring that shows that operating income increased 42 percent within two years of the buyout and cash flow increased 96 percent. The free cash flow doubled within two years of the buyout. And, yet, R&D expenditures increased.

Finally, I would like to comment on our financial institutions and I am just going to quote the statement of the Controller of the Currency, Robert Clarke, about the impact of debt on financial institutions in America. "Our inquiries do not lead us to conclude that national bank participation in leveraged financing posed undue safety and soundness concerns. Instead, we found that when well-managed, well-conceived, such transactions offered banks a business opportunity that was important because of the decline in their traditional commercial financing activities.

"In general, the banks had established and were following well-thought-out policies and philosophies for the leveraged portfolio. Because leveraged financing is a type of financing, not an industry, not all leveraged transactions will be affected in the same way by changes in the economy. The risk of default on such transactions varies from company to company and from industry to industry, but is generally dispersed."

In closing, I would like to comment on one thing and that is that we are going through a period of change in the financial markets, that the inefficient public corporations are under pressure to change, and are being forced to do so. Heretofore, inefficient companies had little impetus, few outside factors to affect them. It is only because these "wild men" now attack them and challenge them, challenge the boards, that people are reexamining the efficiency of what has become an inefficient American public corporation.

MR. EIZENSTAT: We have now gotten the Wall Street view. To get the most out of our day today, it is important to hear your questions and comments to Roger and Michael, so we will throw the floor open for those comments and views. If you have comments, keep them brief because we really want to elicit responses from our panel.

We do want to record your questions and comments for the record. So, please identify yourself. However, if you wish to ask a question incognito, we will allow that, too.

MR. FLAX: I am Al Flax and I am with the National Academy of Engineering.

I have two questions that could be addressed to either of the two speakers and, perhaps, the speaker last night.

Number one is that if this change, as represented by the complex picture of the leveraged buyout, is so critical to the performance of the economy, how does the Japanese economy get along without it? Parenthetically, *Financial Week* says they have a different political and social system, which is not really the answer because we could change ours, too. Something like the Glass-Stegall Act that prevents banks from getting in bed with companies can be changed as easily as it was instituted in the first place. And I was there when it was instituted in the first place.

Now, the second question. I am quite impressed by the argument that takeovers are only of non-R&D companies because there is one capital market and if we bid up that capital market with activities that are lucrative but unproductive in a long-term sense—a pejorative implication here—it is going to keep us out of investments in R&D, and incidentally not just in R&D, but what is more important, commercialization, that improves much more than R&D. It improves capital investment and launches new products.

MR. ALTMAN: Well, Mike Tokarz and I may have different views on this, and I hope so because that will make it more interesting. But I did not intend to say, because I don't believe it, that restructurings are vital to the health and future vitality of American industry; quite the opposite. I just said that restructurings are a consequence of broader savings, investment and cost of capital trends, which are really quite negative for American industry and for the competitiveness of this country and ultimately for the most important goal all of us have, which is rising standards of living for our citizens. So I guess I take issue with the premise of your question as far as my remarks are concerned.

As to the part of your question on Japan, I am not actually sure that the Japanese will not eventually find restructurings more and more widespread in Japan, as their financial system increasingly is reformed and liberalized. You are in some respects beginning to see some of that, as the merger and acquisition trend in Japan is rising, both internally and externally.

I would point out that restructurings are growing quite rapidly in Europe, particularly in the U.K. The Europe 1992 phenomenon, it seems to me, is going to open up more of Europe—at least internally within the boundaries of the EC—to restructurings, which originate from the wider market that is going to be an opportunity for investors.

So, I agree with the premise of your first question, rather than disagree with it.

MR. EIZENSTAT: Roger, it seems to me that the question about why the Japanese get along without it underscores your point, which is—if your point is that these restructurings are a consequence of our capital cost problem, the Japanese don't have the same capital cost problem; therefore, they don't have the same pressures on restructuring. I mean, it is another way of looking at it.

MR. TOKARZ: Well, I, too, think that it would be a misunderstanding of my thoughts on whether or not they are vital. In aggregate, KKR for 14 years has done only 35 total transactions, a small number relative to the entire number of corporations in America. What I find unique, however, in comparing our economic system and our governmental interplay in it with Japan's, is that the capital flows in Japan include no significant syphoning off of such flows for purposes of defense. Clearly, our budget here has a large component of defense. As a matter of fact, in many respects, we are a world guardian.

In terms of how they manage their society, I think we are just at different stages. After World War II, they rebuilt basically from scratch. You can argue about cultural differences, but that culture, that society, had to have long-term views. They were not able to get short-term consumption, period, and the political, governmental and economic systems there, I think, were brilliantly coordinated to gain international world competitiveness over a longer horizon.

Our society at that stage was in a totally different phase and I think they interrelate in ways that are difficult to understand; the science of economics is still quite imperfect. So, I don't think restructurings or buyouts per se are going to solve any problem at all or are going to be vital to American competitiveness. I think, like Roger, that they are by-products of some gross inefficiencies, particularly in the capital markets and particularly in our entire economic integration between government and the private sector.

MR. NOLFI: I am Frank Nolfi, with the Alcan Aluminum Corporation.

My question is mainly directed to Mr. Tokarz.

If I take you at your word that you, in fact, correctly describe the case that buyouts tend to increase the efficiency of corporations, would you also say though that you have dismissed in aggregate manufacturing in the so-called high-tech industries? And if I take you at your word then, I guess I would like to comment, what about the manufacturing industries? You say they have not been successful. If not, why? Do you have any feelings about these industries in aggregate? You really did draw the distinction and in all the cases that you cited as successful examples of buyouts, they were not manufacturing industries.

Is the implication that we are to dismiss these or say that the prudent investment of capital should be away from these industries? I am not sure I completely understand the implications of your remarks.

MR. TOKARZ: My remarks did highlight ones that were not manufacturing; although we do have very successful manufacturing leveraged buyouts in our portfolio, including some that have had to compete vigorously against foreign competition.

I will give you an example of one. As a matter of fact, the first large buyout done with a publicly-held company was Houdaille Industries. Houdaille Industries was formed back in the twenties and it made automotive parts. It grew to be one of the top 500 companies in America. It was very, very stable and 40 percent of its profitability was generated from machine tools; another 40 percent from chrome-plated steel automotive bumpers and the remainder from industrial products.

We had the unfortunate mistiming of purchasing that company in 1979, which was nearly after the peak of industrial activity in the United States. By 1980, the 40 percent of their profitability that was generated through the manufacture of chrome-plated steel bumpers was gone. There were no more chrome-plated steel bumpers, period, bought in the United States. So, we lost 40 percent of our income within two years of buying the company.

Additionally, in this same period of time from 1979 to 1985, our friendly Japanese competitors targeted the machine tool industry—another 40 percent of our profitability—as an area for penetration and market share gain in the United States. Between 1979 and 1985 their share of the U.S. machine tool market grew from 3.6 to 84.5 percent. Of course, that was another 40 percent of this company's profitability.

At the same time, as you may remember, in the early eighties interest rates soared to over 20 percent for the prime rate. It was one of the most adverse interest rate environments that we've had.

Despite the gross adversities vis-à-vis foreign competition, vis-à-vis change in domestic markets, the management, who had a significant ownership in that company, was able to aggressively manage their balance sheet. They took what would be considered write-offs on an accounting basis, that had cash-positive impacts on the corporation, so they could redeploy their assets aggressively and effectively. Not only did they survive what would be catastrophic for most companies in terms of the four-year period, but they came out and flourished, acquiring another company 50 percent of their size and now are actually public on the New York Stock Exchange.

So, to answer your question in a straight-on manner, manufacturing companies are actually very good potential buyout candidates for us, or for restructurings of any type. Typically, the whole purpose of the restructuring is to get ownership in the hands of the managers, who then view it as though it is their house and get that marginal benefit, the marginal productivity that Kaplan from the University of Chicago and others have identified as efficiency gains.

Quite honestly, our studies don't show quite as strong results as theirs do, but, generally speaking, manufacturing companies are attractive buyout candidates.

MR. EIZENSTAT: Michael, you mentioned something to me that you didn't mention. I think it is an interesting figure for everybody to have, that 6 percent figure for ownership shares of buyout by your managers.

MR. TOKARZ: We feel, as I should mention to you, that ownership in the hands of management improves productivity. In our companies, on average, 15 percent of the ownership is turned over to management. They buy 15 percent, relative to a public ownership where management owns less than 1 percent on average.

In terms of CEOs, in our companies on average the CEO owns 6 percent versus a public corporation where the CEO owns less than a quarter of 1 percent.

MR. HOLLAND: My name is Max Holland. [Editor's Note: Mr. Holland is Contributing Editor of *The Nation*.]

I would like to ask Mr. Tokarz some questions about the Houdaille buyout as a success story. The fact is Houdaille doesn't exist anymore as a public corporation. What Mr. Tokarz described as going back on the stock exchange is sort of a lump of Houdaille, Ivex Corporation—20 percent of what Houdaille used to be. He spoke earlier about not selling the garden to neighbors. The fact is the best part of Houdaille, a company called John Kramer, was sold to a British conglomerate in order to retire the debt that Houdaille had acquired.

I would like to ask him how his assertion on capital expenditures applies to the experience of Houdaille. Prior to the buyout, in constant '82 dollars, capital expenditures were on the level of 20, 27, 28 million dollars. That is 4.9, 3.3, and 5.4 as a percentage of sales in constant dollars. After the buyout, of course, figures aren't available, but the most recent ones that have to be supplied to the SEC showed that capital expenditures were 1.4, 1.7, and 2.1 percent of sales.

I would like to ask him to address that question.

MR. TOKARZ: Mr. Holland, I heard more than one question. I am not sure I got them all.

MR. HOLLAND: Does the best part of a company have to be sold to a foreign corporation in order to pay down the debt? Is that success?

MR. TOKARZ: When you say the best part of the corporation was sold to a foreign corporation, the entire Houdaille Industries was sold to that foreign corporation in 1987. That is correct.

MR. HOLLAND: As Mr. Colberg points out in his complaint against KKR, it was obviously a sweetheart deal to sell the rest back to KKR.

MR. TOKARZ: Well, Mr. Colberg does allege that, and Mr. Colberg is going to receive an affidavit from that foreign company shortly, which will indicate that his understanding of that circumstance is actually inaccurate, according to that company's deal. We sold the entire corporation to TI Group of U.K. in 1987. The TI Group, subsequent to the crash, for purposes in and of itself, determined that it wanted to sell what is a group of six companies. They had it on the public markets for sale.

As a matter of fact, Morgan Stanley, I understand, had an arrangement with them to buy those companies. We became aware of that and offered to buy those companies from them and did so and since have taken those companies public.

MR. ALTMAN: Could I just add one thing, which maybe puts this in a bit of perspective? I am no expert on Houdaille, but it is important to understand that leveraged buyouts, by their very essence, at least most of them, are designed to eventually be liquified, as I mentioned in my comments, by the organizer, by the equity investor.

So, the notion of selling all of the company, as happened in Houdaille, is exactly the point of orginally organizing the buyout. Now, we could debate as a separate matter whether that is good or bad, but if you look at buyouts that have been organized over fairly long periods of time or rather early on as Houdaille was, you are going to find that most of them which worked were sold or refinanced substantially. That is the point.

MR. HOLLAND: That is true, but, Mr. Altman, if you were familiar with the Houdaille case, you would realize that the sale came about because Houdaille precisely didn't reach the pot of gold, at which point they could go back to the stock exchange—

MR. ALTMAN: Well, as I said, I am not trying to address the specifics with Houdaille. I am wholly unfamiliar with them. I am just—

MR. HOLLAND: I think we ought to talk about specifics sometime in order to understand the issue.

MR. ALTMAN: I am just trying to point out that it is correct that the core assets are eventually sold but the way to think of that is that it is the basic goal at the beginning, from the point of view of those who organize them.

MR. HOLLAND: That is absolutely true. That is the basic goal, and the financial investors have absolutely no interest in running it for the long term. It is simply that three- to five-year period, after which they go back and reach the pot of gold.

But I would still like Mr. Tokarz to address the issue of capital expenditures at Houdaille as a percentage of sales.

MR. TOKARZ: Well, two things. One is that Houdaille was owned by us far in excess of the three- to five-year period. So, I mean, I would quarrel—

MR. HOLLAND: Why was that?

MR. TOKARZ: Well, for numerous reasons associated with that company's life cycle, growth, redistribution of its assets and reorganization under duress. As I mentioned to you earlier, 40 percent of its business was in steel chrome-plated bumpers, 40 percent was in machine tools. One was totally wiped; the other, obviously, was under some pressure.

That company had to redeploy its assets and it took a longer period of time for it to do well than it might in some other cases. In the case of Houdaille, all of the debt was retired ahead of schedule and the investors had a very fine return, I think around 33 percent, but it took a long period of time. That is 33 percent compounded per annum.

As far as capital expenditures as a percent of sales, I think part and parcel to it is the time frames you use. Because if you take a look at the period 1982 to 1985, industrial sales declined dramatically in the U.S. and as a result, the profitability of those companies declined as well. Therefore, capital spending was curtailed because companies were husbanding their cash in an adverse environment.

Consequently, when things improved and sales were more robust and opportunities for investment of capital generated by the business were more attractive, companies like Houdaille had increased capital investment in their plant and equipment.

MR. HOLLAND: So, you are saying it had nothing to do with the debt?

MR. TOKARZ: No, I think quite honestly, Mr. Holland, the way we structure our debt is critical to the understanding of how monies are spent. Because if you take for the moment the premise that our objective is to increase value over the long term, otherwise you wouldn't get a return, it would be incongruent to take and starve our companies, directly or indirectly, for the capital that they need to become more robust or more valuable.

So, in the case of Houdaille, I think it very well could have been less than it was in other periods of time. At any point in time where capital is constrained or declines are relative to any measure, sales or otherwise, it is because management is determining when it is best to make that investment and under what conditions, to maximize the value in the long run. The whole premise is value appreciation long run.

So, at any—

MR. HOLLAND: Long run being three to five years.

MR. TOKARZ: Well, like I said, in the case of Houdaille, it was something like nine years or eight years that we owned that company. And, by the way, the change of corporate control from us to the TI Group had in no way a negative impact on the employees of that corporation, its health or its future. It was a simple change of ownership, which occurs daily on the stock market: The shares of many of these companies out there change significantly every day; General Motors shares, IBM shares trade all over the place, and it does not affect the basic company.

As a matter of fact, some of the divisions sold from that company and others actually get into more productive hands. I will switch gears for a minute to Beatrice, where we sold the Playtex unit, which makes bras and tampons and owned Max Factor. We did a leveraged buyout and we sold that company to the management.

I was shocked when I saw management's budget for operating profit of that company a year after we sold it to them. When I had the company, Beatrice had $120 million of operating profit, of which $40 million was really Max Factor. They sold Max Factor to Revlon, another company in its business—imagine Beatrice, by the way, owning Max Factor—they sold that business, which had $40 million of operating profit, in other words, one-third of the $120 million, and their budget for that very year after they sold it was $120 million of operating profit. So, they improved the productivity of the remaining businesses 50 percent.

Now, even we, who feel that we have some impact on improved productivity, didn't do as well in that case as the management that we sold it to. So, sometimes these redistributions actually improve productivity even further.

MR. EIZENSTAT: Last comment.

MR. HOLLAND: I would just like to say that I think Mr. Tokarz is offering a disingenuous version of what happened inside Houdaille and inside Houdaille's divisions after the leveraged buyout. I talked to people who worked at different divisions and I don't think that it became a more efficient corporation after the buyout.

MR. TOKARZ: I think then for the audience's benefit that people should know that Mr. Holland wrote a book called, *When the Machine Stopped.* I believe that Mr. Holland's dad worked at one of the divisions of Houdaille. It was the machine tool division and you are welcome to read his book. I think, in part, that is why he is so concerned about Houdaille. Quite honestly I would clearly argue the "disingenuous." On the other hand, I can only submit to you the record that was submitted to the SEC on Houdaille, which, by the way, shows that the company improved its value, retired its debt and met all its obligations over the period of time in question.

MR. EIZENSTAT: I now feel like a judge instead of a moderator.

MS. BLAIR: I am Margaret Blair from the Brookings Institution. I address my question to both speakers.

Mr. Tokarz spoke of the magic of leverage and the efficiency that can be had in an organization when the organization is very closely held, as though this was some sort of universal truth about organizational behavior. Yet Mr. Altman said it is the intention from the beginning to go public in three to five years. How do you reconcile those two facts?

MR. ALTMAN: I think that they are related to each other in the big picture, but they are not directly related, those two points you make.

First of all, just for clarification, I said the goal usually is to liquify the equity investment over a three- to five-year period, or something like that, which, as Mike points out, differs from company to company because of cycles and so on. That doesn't mean to go public with it. A lot of these are refinanced and remain in private hands and there are other ways of doing that.

But the point about management incentives is really about as basic as any, in terms of human behavior. Mike pointed out the very, very basic difference between the public corporation format and the private corporation format, and Professor Jensen has published a piece, which is really very provocative and I commend to all of you, arguing that for a lot of underlying long-term reasons, the day for many corporations where public ownership was the most efficient form of ownership is over. That has to do with access to financing that private companies now have that they didn't used to have, and with the way the capital markets have evolved; it is no longer necessary to be publicly owned, for instance, to finance a business efficiently.

I would say there is quite a bit to that argument that Professor Jensen makes. It seems to me that there has been for many years in the United States and in other industrialized countries, an orientation to reported profit; in other words, earnings per share for financial accounting purposes having been the primary measure of corporate performance. That increasingly has been a very short-term measure, as there has been a focus on quarterly earnings per share and even often on very small differences among quarterly earnings per share, one quarter versus the next, one year's quarter versus a comparable year earlier period, and so forth. You are all familiar with the arguments over that, as to whether it is healthy or unhealthy.

The point is that is not the case in a private ownership format. Usually, the orientation is somewhat longer term. Often, it is a lot longer term and it is often to cash flow return on investment rather than reported earnings. So many accounting factors determine what reported earnings are, which can be, not always, but can be irrelevant in a private format.

So, my answer to you would be the set of incentives that revolve around ownership, but also revolve around just measures of performance, are, in fact, very different in a private format than they are in a public format. They do produce different motivations and have produced in a lot of cases, which have been carefully studied by many people more capable than I, different results.

MR. TOKARZ: I accept that answer.

MR. KLIEM: Peter Kliem, Polaroid Corporation.

Gentlemen, I do have a question for you but before I ask it, let me join you on two premises. The first one is that I join you on the premise that productivity is the central issue that we are dealing with. And I also personally happen to believe that ownership is a powerful motivator in the dilemma that we have.

My question to the panel, particularly you, Mr. Tokarz, is the following: Do you emphasize the value of ownership within management in a corporation? It is my belief that it is the intellectual capital of a corporation that really in the end will change productivity, but by that I do not just mean the intellectual community of R&D, but of all employees and all members.

One vehicle to bring that ownership about, not just within the management ranks, but within all of those who can change that equation of productivity, is employee ownership in a publicly-held corporation. It is my impression that Wall Street in general is not very supportive of that concept. I would be interested in your personalized view of this.

MR. TOKARZ: There is no question in my mind that if the corporation's management and the board of directors deliver to the employees ownership of that company, by whatever means, whether it be through distribution of shares or any other mechanism, those employees will be much more attuned to productivity gains because they have an ownership interest.

In Houdaille, if I might, when we took it public, we gave to each and every employee shares of that company. The reason is just what we have tried to emphasize: When we deliver this 10 to 15 to 20 percent of the company to the management of the enterprise, when we take it public or when we buy it—take it private, I should say—it is distributed down to various ranks in the company. I think in Beatrice it was down to approximately 250 or 300 people.

In some instances we stimulate ownership through delivery of the unit that literally mimics or mirrors the stock ownership that the individual investors have at the higher levels. So, from a Wall Street standpoint, there are so many companies that would be totally inappropriate for a buyout; probably the vast percentage of companies are grossly inappropriate for what we do. I would submit to you that mechanism should be improved, to deliver ownership to the employees, and I favor that totally.

MR. EIZENSTAT: Both of our speakers have to leave at 10:30 a.m. and I want to take this time to ask the last question of the panel.

It is admittedly an untutored question and just based on intuition and I would like both panelists to respond.

It seems to me intuitively that if you have a corporation which is more heavily debt laden as a result of a buyout, that in order to service that debt, it has to divert assets that might have been used for research and development and other things to servicing the debt. Why is that intuitive feeling incorrect and why do some of the studies, which Joe Grundfest made, not seem to back that up?

MR. ALTMAN: Well, first of all, Stu, I agree with you on what seems to be intuitive and I am not persuaded that the entire set of factors that I tried to talk about, of which, I think, restructurings are a consequence or a by-product, aren't really quite negative for research and development. We all know that basically the R&D trends are negative in the United States, in terms of private R&D share of GNP. It is a declining percentage and it is well below that of most of our industrialized competitors, not just Japan and Germany, and there are a lot of reasons for that. But it seems to me more important than any others are these points about negative savings and investment trends and productivity trends.

So I look upon the points that Mr. Framm made in his paper as correct, but only in a very narrow sense. First of all, it is a rather small sample and I think it is just one small slice of the whole restructuring trend (and an even smaller slice of the whole pie) that I think is the most relevant part of it.

Now, on a much different level, asset sales are not central to every buyout in terms of making it work. There are actually a lot of buyouts. Mike could probably quote the statistics much better. There are a lot of buyouts where leverage is applied and improvements are made in the basic operations of the company. Assets, or at least major assets, are not sold; research and development continues either at the prior pace or, as Mike talked about, in some cases is expanded. The improved operating profit of the business, together with the fact that previously it was unleveraged, permit the enterprise to move forward and the equity investors ultimately to prosper without asset sales.

So, it just isn't the case that it is a requirement of every transaction. Also, if you think about Mr. Flamm's measurement techniques, you do have to ask yourself: if assets are sold, what is the ongoing research and development trend within those assets in the hands of the new owner? From the point of view of our society as a whole or our economy as a whole, that is really the germane question.

For instance, one thing that might have been mentioned here this morning, the biggest restructuring going on in the world right now is British American Tobacco, BAT, a huge conglomerate. It has been attacked by a very colorful group of characters, led by Sir James Goldsmith, and has now announced its own restructuring, which essentially is a deconglomeratization of BAT. I don't know a lot about BAT. I am not involved in it. I am not an expert, but it is not difficult to conclude from afar that BAT is in tobacco, insurance, retailing, to pick three main businesses, for no particular reason, other than it was nice to get bigger.

So, it is hard for me to look at that—and I am in a certain sense a practitioner, not involved in that one—it is hard for me to look at that and say isn't it bad that BAT is being restructured. At least there is a good chance that those assets are going to be more productively used in separate hands—retailing if they are sold to other retailers; insurance, in the hands of insurance owners; and tobacco elsewhere—than at present.

MR. EIZENSTAT: Michael, could you explain to me why my intuition about this may be incorrect?

MR. TOKARZ: I think it is natural to assume that if you are leveraged and you must husband your individual dollars and cents, that you are going to deprive other areas where if you had excess cash flow, you would deploy it. I think that is a natural and a logical deduction. Why then does the evidence suggest that it doesn't occur?

Well, what we and Kaplan and others find is that the actual enterprise is 46 percent more profitable within two years and 96 percent more productive in generating cash. By the way, when we design them, we design them based upon very, very modest expectations. We don't plan for a 46 percent increase in operating profit, I can assure you. And again, it is a very small number of companies. I wouldn't want to suggest to you that this is the panacea for American or any other societal change.

But what happens in these companies is that because of the ownership change to the employees and the management, they become more productive and, therefore, they are not short of cash and can continue those programs and, often times, because they are value oriented, look to the longer run.

So, I think it is a result of the result that permits that fear in most instances not to be realized.

MR. EIZENSTAT: I think everyone would agree we have had a provocative panel and an interesting panel. We appreciate both of you coming.

We will break until 10:45 a.m. and we will start right on time at 10:45. (Brief recess.)

MR. EIZENSTAT: We are really particularly fortunate and, indeed, if I may say so to both Mac and Henry, privileged to have two exceptional CEOs who have taken time out from their very busy schedules to participate in our panel on the corporate view, and we do appreciate both of you coming very deeply.

Mac Booth is Chief Executive Officer and President of the Polaroid Corporation, positions he has held since 1986. He has worked for Polaroid since 1962, after his Air Force service, in a variety of positions. Among his many accomplishments was to help develop Polaroid's first color film, the SX-70 Land Film. He knows every aspect of Polaroid, from its research labs to its management headquarters.

Mr. Booth has also made corporate history during the last two years by being among the first CEOs to maintain his company's independence in the face of a hostile takeover, without paying greenmail and without radically changing the structure of the company. I am sure he will be telling us more about that experience, as well as its significance to our central topic of restructuring and research and development.

Mr. Booth holds a B.S. in mechanical engineering from Cornell and also an MBA from Cornell. Mac, we appreciate you coming.

MR. BOOTH: Good morning, everyone. I am very pleased to be here this morning, to be a part of a debate, which I know we have already seen part of and I am sure we will see more of as we go along, that is an extremely important issue in my sense of what is facing American life today.

I think that either one of the benefits or one of the problems of coming on after some other people have had an opportunity to say their thing is that you want to rebut immediately what you have heard. But before I do start my formal talk, let me just try to say a few things that I have heard last night and had some opportunity to think about, obviously, over the evening and here this morning.

I personally don't think the question here is good guys and bad guys and I think if we polarize it in that way, we are doing the wrong thing. The question that we should be discussing today is the impact, and what I consider a significant impact, of high leverage on R&D, no matter who is running the company, no matter who is controlling it. I think that is the question that is facing us.

As Peter Kliem said, you won't find any person in the Polaroid management questioning ownership. We believe in it. We have gone through a very serious ownership activity in the company, where 20 percent of our company's ownership is held by the employees, which they have paid for out of their wages. It has been deducted from my wage and from Peter's wage and from all of our wages. We are paying for that stock ownership. We believe in it. We believe it is important.

We believe this country would be much better off if we had more ownership by the employees. So, I don't think that is a question that we should be arguing. I think the question that you might ask is, is going private the only way to get ownership? I obviously believe it is not and I don't believe that private ownership is the long-term benefit for this country. I believe public scrutiny is important and I believe there should continue to be public scrutiny for public corporations. I believe it keeps you healthy and it keeps you on your toes and it keeps forcing you to really justify why you are doing things.

So, as I say, I think that the question of high leverage is the issue. Whether it is restructuring, LBOs, however it is done, what is its effect on R&D? So, I believe that this broad issue that we are facing is how do we make sure that U.S. industry keeps or, better yet, improves its worldwide competitive position and, thus, provides a handsome return for all its shareholders and a growing standard of living for all Americans.

There are many factors that contribute to our relative worldwide competitive position. Education of our youth, we have heard discussions of that. Education of our continuing work force, I don't believe there is enough emphasis on that. National fiscal policy, such as tax and savings incentives; antitrust policies; uses of national financial funding for research and development; and of course what we are here today to discuss is corporate research and development. I believe everyone here would agree that corporate research and development is vitally important to the long-term success of our corporations.

The three issues that we need to be discussing are: One, are we getting our money's worth from our corporate research and development expenses? Two, what is the effect on research of high debt and the large cash flow requirements to pay off that debt? And, three, what is the danger that the desire for unusually high annual rates of return on investment, of 30 percent or more, will negatively impact our corporate research and development programs?

It is my belief that such demands will slowly destroy U.S. firms, by making them uncompetitive in an increasingly competitive worldwide market. "Slowly destroy U.S. firms" is a strong statement. Unfortunately, it may be an understatement. The destruction may come more quickly than slowly.

There is no question in my mind that without invention and innovation, we will become a second-class economy. And, therefore, we must adequately fund and adequately manage corporate research and development activities or we will eventually become a nation of followers rather than leaders.

There is no mystery in what we are talking about today. Research and development expenditures are not intended to increase sales or profits in the years in which those expenditures are incurred. In fact, they reduce profits and they reduce cash flow in those years. Heavily leveraged companies require large cash flows to meet the huge interest payments on their increased debts, and in situations where a public corporation is taken private, not only is there usually a large debt payment, there is also a need to make the company appear profitable. So, when it is again taken public, usually in three to five years, the financial reward for the private owner is huge, which is the principal motivator.

Under such circumstances, research and development expenditures sit on the profit and loss sheet like a fat plum ready to be picked. R&D becomes a predictable target for reduced spending. Of course, neither the corporate raider nor the leveraged buyout team will announce plans to drastically cut or even eliminate research and development expenditures in order to dramatically increase the cash flow of the company.

You won't hear any of them say, "I don't need to spend much money on research and development because I won't be around in three to five years to reap the benefits of that research." Rather, the words that are used are more like this: Look at the money that is being spent on research. Are the shareholders really getting their money's worth? Are they getting the kind of efficient return they deserve? Where is research leading this company?

These are all good questions; in fact, the same questions that all good companies ask themselves on a regular basis, but from a different perspective. The questions we ask sound like this: How can we use our research capabilities more productively? How can we get twice as much from our existing resources? How important is the role of research in our vision of the future and what is the potential return for our shareholders now and in the future?

Research becomes one of the three or four focal points of attention by all parties in a hostile takeover or leveraged buyout situation. Everyone takes a long, hard look at research. That look focuses discussion about long- and short-term strategies, product development, focusing of resources, et cetera. And this dialogue is often public in nature.

I would like to offer Polaroid's recent experience to illustrate the nature of this dialogue, but before I do, I think some historical perspective might be helpful.

From the time of Polaroid's founding in 1937, Edwin Land set out to build a great enterprise, based on a vision of what a research-driven company could accomplish for its customers, employees, communities and society. It was a grand vision that outlined very specifically the significant role of research in the company's organization and in its culture, and it served as the catalyst for our growth.

Dr. Land built a new field from the basic research conducted by himself and his fellow associates. He relied on his tremendous faith that the cumulative benefits of research would eventually produce products and processes of great appeal and value. He believed quite clearly that his company should be as much a research organization as it is a manufacturing company; fiercely creative in its pure science, its contributions comparable with those of university laboratories.

Dr. Land's innate belief in the value of research reflected the times as well, especially as postwar America prepared to change from a defense-oriented economy to a domestic business boom. It was a wonderful and optimistic strategy that was instrumental to American business development from 1945 to the mid to late seventies.

In 1945, Vannevar Bush, FDR's science adviser, outlined a vision of the role of research in postwar America. I would like to quote from him. "Basic research leads to new knowledge. It provides scientific capital. It creates the fund from which the practical applications of knowledge must be drawn. New products and new processes do not appear full grown. They are founded on new principles and new conceptions, which, in turn, are painstakingly developed by research in the purest realms of science."

He continued, "Today, it is truer than ever that basic research is the pacemaker of technological progress. In the nineteenth century, Yankee mechanical ingenuity, building largely upon the basic discoveries of European scientists, could greatly advance the technical arts. Now, the situation is different. A nation which depends upon others for its new basic scientific knowledge will be slow in its industrial progress and weak in its competitive position in world trade, regardless of mechanical skill."

This view was echoed by leaders at M.I.T., Cal Tech, and the Carnegie Institution, the National Science Foundation and other members of the business and science establishment. Science and technology had in a major way won the war and science would now pave the way for a great American business resurgence in peacetime.

Vannevar Bush and Edwin Land had never heard of a leveraged buyout. In fact, they had only one real mission in life and that was to build our economy through invention and innovation.

Polaroid's early success proved to a model for other postwar start-ups to emulate. Yet, as Polaroid grew from a small enterprise to a $2 billion business, the role of research evolved, changing and modifying its place in the conduct of our business. World-class marketing, engineering and manufacturing capabilities grew and developed the kind of balanced weight one would expect in a large corporation.

Still, at each stage of our development, new products led the way to growth, and each new product could find its roots somewhere in the basic building blocks developed by our scientists and engineers in our laboratories.

The world marketplace also evolved, as did the development of complex imaging-related technologies. By the late seventies, we realized that to compete we could not solely grow our own, that to invest and become proficient in every complex technology needed for Polaroid's future would be too expensive and too time-consuming. We realized that we had to bring our marketing intelligence and skills earlier into the product development cycle and that our engineering had to be more closely coordinated with our photochemical research.

We had to find new and different ways to accomplish our task, new and different ways to approach product development and research in general. We were well on our way to implementing these changes in the way that we approach our jobs within Polaroid, when in July 1988, Shamrock Holdings began what was to become a nine-month battle for the control of the Polaroid Corporation.

Needless to say, I am very pleased to have the opportunity to be here in my present position to tell you some of the facts of that experience. The ensuing battle for Polaroid in the courts and the press, in the hearts and minds of our shareholders, has been well documented. I will save you the blow-by-blow account for another time, unless you want to ask me some of the particulars in the question and answer period.

More important, though, I believe, for this forum is the focus on research. That became a major factor in the public debate about the future of Polaroid. Indeed, at the same time that our research activities were being attacked externally by Shamrock Holdings, they were being examined and reexamined in great detail within the corporation itself.

In Polaroid, the takeover threat was the ultimate confrontation with the future and the rapidly changing world we live in. The future had landed on our doorstep. The questions that we had to ask ourselves about our research and development activities reflected the urgency of the situation. We asked, what are our expectations really? Are they realistic? Are they worth it? Does the potential market justify the research costs? What can we pare down? What can we streamline? Is there a need to divest some projects that simply cannot be justified? Where in our research and development activities were we not getting our money's worth? How can we get the job done without expanding our resources? How can we conduct our research more passionately in new creative ways, faster and more productively?

These are not trivial questions in any business climate. The accelerated pace of the takeover environment made our inquiries all the more intense. Under the takeover threat, our version of the future of Polaroid became a very key factor. That future was greatly dependent on our R&D activity.

We knew this more instinctively than we did objectively. We also knew that the world, both inside and outside the company, needed to hear our vision stated in concrete terms. In fact, what started as a need to quickly communicate to the outside world became an internal source of energy.

Managements know that investors may believe in the importance of R&D, but if presented with a handsome, all-cash premium to the market price of their stock, they tend to forget about the long-term value. Our job then became a challenge of communicating to the shareholder community the value of what is going on in our laboratories. If the market correctly valued the potential found in the company labs, then the shareholders facing the decision as to what to do with their shares would have a better basis for making that decision. That challenge continues today.

In its study, *Made in America,* the M.I.T. Commission on Industrial Productivity noted that managements may have a market-imposed tendency to overdiscount R&D in the option values arising out of R&D. Although the Commission and the study it drew from relate this tendency to investment decision analysis, it should also apply to the valuation models used by institutions and analysts in reviewing corporate performance and market value. This would suggest to me the development of a new evaluation system for use in growing, technically-oriented businesses.

Today, America remains a leader in industrial research, but the world has changed. In the past, there wasn't any debate on the depth or breadth of research performed in the United States. Today, there is constant debate over whether we are falling behind or how fast we are falling behind in commercializing the technology being developed in our laboratories.

There is no question in my mind that we are slipping and I strongly believe that very high debt and the consequent need for large cash flows foster a significant short-term emphasis in business and industry, at the expense of important R&D projects, and that this is a strong negative force on our collective future.

That doesn't belie the fact that we must be better in our approach to research. I don't think that the raiders and the leveraged buyout kingpins are wrong when they demand that corporations look at their R&D spending. If there are lasting lessons to be learned from the Polaroid experience, they relate to how we conduct our research, as much as what we choose to invest in.

Let me suggest what some of these lessons are. Research and development cannot function productively in splendid isolation. It must be connected to the business environment in very direct ways. Researchers must be part of an integrated team of product developers that includes manufacturing and marketing personnel. They must understand the competitive environment in which we live, the realities of the marketplace and how their work relates to the financial and product strategies of the corporation.

They must also understand the need to abandon some projects of truly brilliant technical brains that just don't fit into the overall corporate strategy, or for which the potential return would be too long in coming. And research in engineering must become as innovative in how it performs its research as it has been in what it delivers. We must find ways of improving the productivity, the effectiveness of research, just as we have improved the productivity of manufacturing.

Research and development will continue to be the driving force for our country's growth, as it has been for the past 40 years, but our owners have a right to expect that we improve the effectiveness of the dollars we are spending. The answer isn't to spend less. The answer must be to spend what we have more effectively, striving for measurements, striving for collaboration and striving for some tough, hard decisions.

If we expect our shareholders to think beyond the short term, then we must share with them the vision that undergirds our research efforts, and we must establish a serious dialogue with the investment community. Major shareholders and influential analysts must understand the potential value of the technologies under development in the laboratory.

They must have a comprehensible blueprint of how technology responds to and capitalizes on the marketplace. It is not necessary to disclose competitively sensitive information, but investors must have enough information to understand the critical option value several years old. Only then will the market value more accurately reflect long-term potential. I can tell you from experience that this dialogue can be a productive experience. Hearing outsiders' perspectives on your company can give you new ideas, keeping your outlook fresh and responsive to that outside world.

This is not a time for excuse-making and hand-wringing in American business. We don't need to protect R&D by keeping it locked in an ivory tower. We don't need to keep R&D isolated from the sweaty reality of the marketplace. We need to demand higher levels of productivity and effectiveness and we need to expose it to the full glare of public scrutiny and we need to sell it.

Thank you very much and I am looking forward to the questions.

MR. EIZENSTAT: Thank you, Mac.

Henry Wendt is Chairman of the Board of SmithKline Beecham. He joined SmithKline in 1955, shortly after receiving his B.A. degree in American diplomatic history from Princeton. He has served in a variety of executive marketing positions in the U.S., Canada and Japan before being elected as president in April of 1976, a position he held until he was selected as chairman in April 1987.

He was named Chief Executive Officer of SmithKline in 1982. He is a member of the Board of Directors of ARCO, the Pharmaceutical Manufacturers Association and the Japan Society. He is chairman of the U.S.-Japan Business Council and a member of the Advisory Council of the Department of East Asian Studies at Princeton. He is also a member of the Business Roundtable and serves on the Roundtable's International Trade and Investment Task Force.

Perhaps as a result of his international outlook, last year he led SmithKline into a merger with a British corporation, Beecham. He will be describing that decision and its significance for us.

Henry, thanks for coming.

MR. WENDT: Thank you, Stu. Good morning, everyone. It is a great pleasure for me to be here. I have been fascinated by the discussion this morning and it is an honor to participate in that with you. It is especially an honor to serve on the same panel as Mac Booth. I admire his spirited fight and all that he has done in his leadership of Polaroid.

I would like to begin this morning with an important confession. I arrived here in Washington early this morning on a company plane. It is a 10-year-old Cessna Twin. It is not very comfortable. It charges out on a fully-loaded basis at 100 dollars an hour. So, with two of us on board, it is definitely competitive on a fare basis to Trump, and it is a darn sight more reliable.

Of course, our topic is of major importance. That is why we are all here; the effect of restructuring on research and development. Goethe said many years ago that the only way to know anything perfectly is to do it. As the chairman of a newly-merged company deeply involved in restructuring, as Stu pointed out, I suppose I fit Goethe's definition.

At SmithKline Beecham, we are definitely doing it. What I have to say then is based on the experience of one company. But in describing that experience perhaps I am also discussing an industry; in our case, the makers of prescription and non-prescription medicines, where research and development is not only a way of life, but the only way of life. Drug companies either innovate or perish.

My experiences and observations may not be typical of all industry; yet, of course, I believe, they are relevant to the topic. Although many take a negative view of the effect of restructuring on research and development, I come down on the positive side.

But I have no doubt that there is also a negative side. This conference was called, if I judge correctly, because it appears that the ground has shifted in American research and development and, indeed, all the previous speakers have reaffirmed that observation. There are two principal facts pointing in that direction: The growth rate of spending on R&D has been slowing measurably since 1984; and U.S. research managers, some of them anyway, observe that corporate restructuring, particularly those restructuring events that involve a great deal of financial leverage, have caused R&D cutbacks at their own and other companies.

I will attempt to make the following points to you this morning. First, I will characterize SmithKline Beecham's status in the world of corporations and describe our recent merger. Then I will discuss the transnational corporation and in doing so, introduce another element into our debate here this morning; that is, of world markets and of national ownership of research. In doing so, I will grapple with the question of whether transnational companies have a good or bad effect upon national economies.

Finally, I will give you my own conclusion, that restructuring under proper conditions is not only no threat to R&D, but, in fact, can give it new life, not only through increased investment but also through the stimuli arising from sheer challenge and change. I will also stress the fact that the American short-term investment horizon, more than any amount of restructuring, is the real culprit causing a slowdown in investment in research and development within the corporate sector.

First, SmithKline Beecham. What is it and how did it get to be that way? It is, of course, a research-intensive, indeed, a technology-intensive corporation, in contrast to at least some of the companies characterized in this morning's discussion as the ideal targets for leveraged buyouts, which are perhaps the most dramatic form of corporate restructuring.

SmithKline Beecham is a direct product of corporate restructuring; in our case, an equity merger, an old-fashioned equity merger, with a strong like-minded partner on a friendly basis, designed to form a more vigorous world competitor. Our prosperity depends entirely on the development of innovative products through technological advances. But there is a new message in the merger of SmithKline and the Beecham group to form SmithKline Beecham. That message is this. We believe we are a member of a new breed. We have joined the ranks of the global or transnational company, which I personally believe will be the major organizational style for many businesses in the next century.

Who are we? How are we different and why are we devoting a half billion dollars, and growing each year, to research and development? And although we are not solely an American company, why is our success good for America?

SmithKline Beecham, with nearly $7 billion in annual sales, is among the top five global companies in prescription pharmaceuticals, over-the-counter medicines and animal health products. We market some of the most prescribed medicines, such as Tagamet for ulcers, and a full line of hospital antibiotics. We market consumer brands that are familiar names to you and other consumers worldwide; Tums, Contac, Sucrets, Sominex, Aquafresh toothpaste, to mention only a few.

How are we different? The fundamental nature of our restructuring is different. Ours is a friendly and purposeful merger of an American company and a British company. This fact can pose a conundrum for Americans, who love to keep score. I might say parenthetically, the British do just as much.

For example, how do you score us if we succeed against Japanese competitors? Is that a win for the American side? Likewise, if we prosper in Italy, does that strengthen America's balance of trade? In other words, when an American company and, in this case, a European company merge, whose side are they on? The answer is that the question is wrong. The question posits a business outlook that belongs to the past. The future business outlook, I believe, will be quite different.

It will perceive a world not so much of national corporations reflecting the aims and ambitions of one country, but a world of true transnational corporations, configured in their nature by the needs of all countries in which they do business. That is a substantial change in market perspective and it does, as I will attempt to show you, have relevance to the future of R&D.

So scorekeeping in the traditional sense may no longer be appropriate. The game has changed and government statisticians are going to find more and more companies that do not fit their traditional categories, because more and more companies are going to become and certainly think transnationally.

SmithKline Beecham reflects a new reality, the reality that technology flows across borders and oceans and land masses in spite of executive fiat or any kind of protectionist legislation. We have all anticipated and for years talked about one world. Now one world has become more than an abstraction. It is a market reality and it is not yet, of course—and perhaps never will be—a political reality, but it is certainly an economic reality. The market reality of the next century will definitely be one world. The EEC's Common Market of 1992 is merely one symbol of a step in that direction.

But we confront not only a market reality, we must accept a technological reality. Patent applications and grants show how things are going. Our competitors—and, again the Japanese are a notable example—patent their discoveries in the United States and in other countries, as well as their homeland. SmithKline Beecham, likewise, patents our discoveries in all the attractive marketplaces, including Japan. Where there is no reasonable patent system, however, we usually don't do much business.

It must be said that the reality of the transnational corporation, of open markets and an international trading system, has not always been seen as good. It has, in fact, been feared in the past and in some quarters it is feared today. But it is precisely that international open trading system that has allowed the non-communist nations to grow and prosper for 30 years, to the great and lasting benefit of all of us who participated and to the very severe detriment of those who dealt themselves out of it. This is the system on a global basis in which SmithKline Beecham must compete.

We were not entirely breaking new ground when we merged with a British company. In the past, as SmithKline & French Laboratories, we conducted much of our primary pharmaceutical research in Great Britain, as well as in the United States. We did so by choice, not by chance. There is a great tradition of medical discovery in the English-speaking world. Many of the quantum leaps from the understanding of how blood circulates, to polio vaccine, to the double helix and the breaking of the DNA code were the work of U.K. scientists, often in collaboration with U.S. scientists.

Dr. James Black, for example, was awarded the Nobel Prize this year for his investigations, in the United Kingdom laboratories of SmithKline, that led to Tagamet, our leading product. So, we are not awakening to the elegance of British science. We are simply institutionalizing our faith in British science and American science.

The R&D perspective for our company is this. As SmithKline Beecham, we will be a larger, stronger investor in science and in innovation than either company could be alone. And, I may say parenthetically, a more efficient one as well.

Moreover, as a global company, we have many scientific and technology-based alliances, with Boehringer Mannheim in Germany, with Suntory in Japan, with Novo-Nordisk, with Nova Pharmaceuticals in Baltimore and with Stanford and Oxford universities. In some cases these are collaborations with individual scientists, although all of those that I just mentioned are institutionalized as broad-based contractual collaborations.

Sometimes we collaborate with a small company with superb technology that is seeking a partner. In every case, whether following our own avenues of R&D or joining with partners, we are matching our needs to the realities of the global market. That means an expanding R&D budget, not a contracting one. And matching our R&D to the global market is one of the keys to understanding SmithKline Beecham and our approach.

The price of admission to the world league of pharmaceutical discovery is an annual R&D investment now of at least a half a billion dollars. We and a handful of other companies are at that level. But if, as I see it, the handful of transnational companies now in existence are prototypes for the future, scientists working in research and development have no reason for concern. Our real concern is whether we will have enough scientists in the next century to do the work that needs to be done.

The stakes are very high for research-oriented pharmaceutical companies. World sales of pharmaceuticals last year totalled $154 billion. Europe led with $43 billion, followed by the U.S. at $39 billion, with Japan coming on strong at $31 billion.

What about the question of national advantage that I mentioned earlier? When SmithKline Beecham plays in the big global league, competing with many U.S. and international companies and cooperating with a few, is that good for America? I believe it is. If we are at the forefront of knowledge, medical advances are made available to U.S. citizens rapidly and efficiently.

We employ the best scientific talent and invest in plant and equipment in the United States, as well as in Britain and other countries. That is good for the employment of scientists and for capital investment. We pay at least as much in taxes as we did as separate companies. With the strength and stability of our greater resources, we will probably pay more in the future. We continue our full program of contributions and other aspects of corporate citizenship.

All these benefits accrue to the United States, although strictly speaking we are not an American company. But what about national goals? Obviously, our goals stand in contrast to the current enthusiasm for national competition—technonationalism, as it has been labeled by some of my Japanese friends—through government-financed consortia.

I realize that some of these efforts have their legitimate origin in the concern over domination of an industrial segment, particularly those deemed vital to defense or national security. That argument is often related to government-backed Japanese or European interests. But the response often seems to me to have an unrealistic, "Fortress America" quality to it; a belief that if we can simply erect the right fence, we will somehow be safe behind our borders. If in doing so we can just outwit the crafty bureaucrats of Japan and Germany, then millions of their consumers will somehow buy our products.

The truth is otherwise. America is not going to compete successfully in other home markets abroad, unless we are prepared to invest time and sweat to know the customer and learn from the customer. For example, we made our initial investment in Japan 23 years ago. We formed partnerships. We have successful businesses there today. We contribute, in fact, to the American positive balance of trade with respect to Japan, but in Japan, as elsewhere, we know that we must continue to build, to learn, and, most of all, to adapt.

*The Economist* magazine commented on the issue of national consortia a few months ago in discussing high-definition television. It pointed out that an American-only consortium of computer and telephonic companies, particularly one financed by the military, hardly seems best suited to develop hot-selling consumer goods, like high-definition TV. Perhaps such efforts will succeed, but I fear for any national consortium that does not fully appreciate the global marketplace.

My conclusion is that the observed dip in R&D funding may be a temporary phenomenon. It is, to be sure, in large part related to the takeover frenzy, but what I believe is this. The long-term tendency for large corporate organizations is to grow into or toward the transnational model. Their imperatives are to meet their markets worldwide and to meet their competitors, who increasingly come from abroad.

That being the case, we can, for the long term, expect increased R&D activity, not less. But there are some corollaries to those propositions. Corporate restructuring on balance is a good thing, regardless of whether it is driven by takeover or a merger. It is a major instrument of economic regeneration. The corporate model for non-restructuring, if you will, is state-owned enterprises. They soon become dedicated to the status quo and unless heavily subsidized are, by definition, non-competitive in the world market; hence, the current trend to privatization in Britain, France, and, now, even some communist countries.

Secondly, restructuring has not hindered research and development funding nearly as much as the short-term investment horizon, tied perhaps, as has been pointed out by Roger Altman this morning and Joe Grundfest last night, to the disadvantageous cost of capital in the United States.

Nevertheless, I am convinced that the short-term investment horizon does demand emphasis on quarter-to-quarter earnings and forces corporations to think short term, while R&D, by its nature, is a very long-term process. The American financial markets are geared to short-term investment. Indeed, on any given day, I think it is fair to say for a publicly-traded New York Stock Exchange corporation, long-term investors are in the minority among the total group of shareholders.

At any rate, the corporate managerial response is geared to producing short-term results. The effect on decisions about R&D investments is, therefore, often predictable and in my judgment is the chief cause for the slowdown in private sector R&D funding in recent years.

The Administration's suggestion for a cut in capital gains tax relates to this topic. Although blocked by Congress, the proposed tax reduction, in my view, was not likely to promote long-term investment in American corporations. On the contrary, the proposals and, indeed, the debate all seem directed to draw even more speculative funds, looking for a quick kill, all of which is the bane of the American economy at present and an obstacle in the way of long-term funding of our R&D.

Finally, an industry, certainly one such as ours, and Mac has already made the point with respect to Polaroid, is as prosperous as its technology, which means that all industry, and especially transnationals, cannot survive without science and scientists. Research and development will inevitably thrive under the stimulus of well-funded, large, fiercely competitive transnational companies.

Now, I look forward very much to the discussion. I am sure between the two of us, we have provoked a few questions. We look forward to the debate.

Thank you.

MR. EIZENSTAT: I wanted to personally thank both Henry and Mac not only for interesting talks, but for talking very directly to the topic, and it seems to me that your different experiences in restructuring have very much enriched the discussion. I hope that this will provoke some questions and that you won't feel in any way intimidated because they are CEOs.

MR. WENDT: No one else does.

MR. EIZENSTAT: Yes—again, for those who may not have been here at the beginning, we are trying to use the mikes. If you will identify yourselves, so we can get you recorded.

MR. BURTON: Yes. I am Stan Burton, Council on Competitiveness. I have a question for Mr. Booth.

In your comments, you talked about the need to look at the productivity, efficiency, effectiveness of R&D within corporations; get R&D closer to the sweaty realities of the marketplace is the way you put it. Do you think your experience with Shamrock and the attempted takeover may have had a healthy impact on the way that Polaroid looked at the whole R&D process, in that it forced you to look at productivity and effectiveness and efficiency of the R&D process within the corporation?

MR. BOOTH: I have been asked that question over not just R&D, but over a number of things that relate to the company. Were we just reacting, if you will, and, therefore, isn't that better, and God love all those guys out there who do that to you.

I have had the chance to testify on that subject in the courtrooms. I said in the court in Delaware that there was no question that the Shamrock phone calls and letters were a bit of a cold shower. It surely got our attention and woke us up. I think that while many people have trouble believing this, it is true. I can assure you that we started much of this long before Shamrock showed up, but as I was saying to Stuart Eisenstat last night, one of the things that made me very upset with that phone call and the subsequent activity is not Stan Gold and Shamrock, but myself, that I didn't move more quickly before the call came, because we knew what we should be doing.

It did act as a stimulus. There is no question. It did not provoke us into doing anything that we didn't have planned. It accelerated it. Now, it is easy for me to sit here now, as still the public company, and talk about this effect. The question is, what if they had won and we had lost, what would the company be like today?

I believe the company would be dramatically different. I believe that was stated in the courtrooms and other places, that research would have been one-third of what it is today in many aspects. We aren't a company which had businesses that could be separated and sold off. We were going to be consolidated and sold in pieces to various activities, in my judgment. That is a personal judgment, not something that necessarily is accurate.

So, I ramble a little bit, but let me just say that there is no question that there was a stimulus there. What I would like to leave for this group and any group I can talk to is, for God's sake, don't wait until the phone call comes. Do it; grab the bull by the horns and look at research. I think it is so important to the future of our country and to our well-being and our standard of living.

I do think because of the history of the forties and fifties and sixties of this wonderful research engine, building our economy, that it was sort of left alone and thought of as an ivory tower. I believe today that you can't think of it that way. It has to react to the sweaty reality of the marketplace and it has to be looked at hard and looked at in a productive way, just like we looked at manufacturing or whatever. So, let's get on top of it before you get behind it and someone is knocking at your door. I don't believe the marketplace does value research, as some of our predecessors have said here.

So, I think you have to be out there knocking at the door and showing shareholders what you are doing. That is difficult and troublesome, because they aren't all friendly when you are doing that, but I think they need to understand what you are doing and what the value of that is. Then, heavens, if someone offers them more money, I guess that is the name of the game and they can take it, but they need to be informed and they need to understand. And that is really what I was trying to say earlier.

MR. WENDT: I might add a comment to that. I don't think the markets value research either. They certainly do not value input or allocation in an investment sense to research. They value research when they see the output on the threshold or already across the threshold of commercialization, after it is all done, in other words.

I, too, believe that the general atmosphere of corporate restructuring, takeovers, potential takeovers, brings a true sobriety to the board room, particularly among the external directors. That is all to the good in my view. That is one reason why I think it is very healthy as a general phenomenon.

But research is extremely fragile. One can point to takeovers where the intent genuinely has been to maintain research, but the very act of takeover has destroyed it. The change in management, managerial climate, the culture of the company is too unsettling to people who are in high demand; they can easily go elsewhere, and they do. There are lots of examples of that. So, just the managerial challenge on the takeover environment of research is extremely difficult and should never be underestimated, regardless of intent.

MR. JARRELL: My name is Gregg Jarrel and I am a professor at the University of Rochester. I used to be the chief economist at the SEC, where we got very much involved in this stuff. I am a speaker this afternoon and I need your help in getting some of my comments prepared.

First, Mr. Wendt, you are obviously in an industry that has probably the heaviest research and development expenditures of almost any industry in the world. Do you think currently or recently that your stock has been undervalued by the U.S. stock market as a result of your R&D expenditures?

MR. WENDT: Yes.

MR. JARRELL: That was easy. I would ask you by approximately how much?

MR. WENDT: Are you working for KKR?

MR. JARRELL: Mr. Wendt, that was an excellent answer. Here is a tougher question. You say that restructuring is basically a healthy force, but you also strongly believe in myopia in the stock market. In the ivory towers and among the policymakers, there is something inconsistent about those two views. Myopia—the problem of a stockmarket that does not appreciate long-term research and development, that discounts it, overdiscounts it—is precisely the kind of a condition that will lead to excessive restructuring, will lead to a substitution of debt for equity, will lead to privatization, leveraged buyouts and a sacrificing of research and development. So, I am having a little trouble—did I hear you right or are you comfortable saying that most of the restructuring activity in the U.S. is a healthy sign and is useful, and at the same time you are comfortable believing that the stockmarket has a fundamental problem of myopia?

MR. WENDT: I am very grateful that you asked that question because obviously I did not make myself very clear. So, now, I will give you a longer response, if I may. I believe restructuring is a healthy process of economic regeneration and when I use the word "restructuring," I do so in the very broadest sense, including many voluntary forms of restructuring initiated by companies at various levels and on a continuous basis, but not excluding mergers, takeovers, or LBOs, either. I think they have on balance provided an overall helpful stimulus.

If restructurings were somehow outlawed by your former employer, I think that over time that would be an unhealthy event. So, I don't think restructuring is the villain of the piece. I do think a short-term investment horizon, which you referred to as myopia, is, indeed, the villain of the piece. Clearly, we must have, all of us, the opportunity to restructure the economic engines. Change is the universal law of life. Anything that prevents change is counterproductive in my view.

I am very, very concerned about the short-term investment horizon in this country and I accept the arguments that we heard this morning relating that to the cost of capital, but there are psychological elements as well and it is not just cost of capital.

I relate it directly to the fact that for large, let's say Fortune 200, corporations listed on the New York Stock Exchange, we see the phenomenon now of every year a turnover of 55 percent of our shareholders. So, we lose more than half of our shareholders in a 12-month period. I think that, therefore, in a two-year period—I can't absolutely prove this, but I think it is quite easy and safe to say—that we will have a continuing shareholder base that represents perhaps a quarter to a third of our shares. The rest are not continuing; they are in and out. In fact, I believe the average holding time, coming at it in another direction, is 4 months and 20 days.

They are not investors; they are not really shareholders. They are not interested in the quality of R&D or what the long-term future for this company is as we move into the 1990s. It is peculiar and special in my view in the United States, and not found to nearly the same degree abroad. That puts enormous pressure on our boards, on the directors, who represent all the shareholders and sit around the board room once a month, viewing the company's strategic plans and operations and investment proposals, and on the management as well.

So, you did, indeed, misunderstand me totally.

MR. JARRELL: Do you favor the tax that has been bandied about in Washington, the special tax on short-term holdings, a transactions-type tax?

MR. WENDT: Yes. I am in favor of what I call a progressive capital gains tax. For example, I believe that holdings for one year or less should be taxed at a higher rate than ordinary income; let's say, just to keep it simple, 50 percent. A few of my friends on Wall Street don't like this, but I really believe this and that if they are held for two years, perhaps it can come down close to the ordinary rate; in three years lower, in four years and maybe in five years zero. That would change this very climate we are discussing.

If we add to that, as Nancy Kassebaum has proposed, a tax on pension funds for their transactions, since they are moving all the shares anyway, then we might really change the climate. Another topic, but it does relate to investment in R&D.

MR. DINNEEN: I am Gerry Dinneen with the National Academy of Engineering.

My question is for Mr. Wendt. It has to do with the growth of transnational corporations and national advantage. Let me define national advantage as an increased standard of living for our people, which says that we should continue to attract manufacturing and also R&D, whether they be from U.S. corporations or transnational. Do you see a role for our government in trying to make the environment better here to attract that kind of investment, whether it comes from U.S. or transnational corporations?

MR. WENDT: Well, that is a very good question. I think that the single best thing the government could do is to change and try to affect the investment climate as just discussed. I think that using tax policy to encourage R&D in this country, rather than discouraging it, would be the next best thing. Perhaps more fundamental than anything, as Mac pointed out very early in his remarks, is the quality of education and, therefore, the quality of the scientists and engineers, and it can't be attacked quickly or with tax policy.

But I have said on other occasions—and I don't mean to pick on Ronald Reagan, but it is easier, I suppose—that if education, particularly in the harder sciences, had received the same priority, not necessarily the same allocation of resources, as the 600-ship Navy, now, in 1989, this would be a very different country. This whole discussion would be different. I mean, if that had started out as a priority of the same order, this morning's meeting would be significantly different.

So, that is where it begins, I think.

MR. EIZENSTAT: Mac, would you like to respond to that in terms of government policy.

MR. BOOTH: I think that I would just be echoing what Henry said here. I do think that there needs to be an incentive for research. I really do believe that and I think that maybe one of the better things is, in general, stay out of it and let the natural things happen. I am in favor of letting Japanese firms build plants and do research and come over here and join the fray. I think that is positive. We sure thought it was positive for many years when we did it that way. I don't know why it still isn't positive.

MR. HARDIS: I am Steve Hardis, Eaton Corporation. I am not on this afternoon's program, so I am going to indulge myself by making a comment. This is my only shot.

I would like to suggest that the speaker last night and the first panel this morning focused on different situations than did these two gentlemen. The data that were commented on last night and this morning draws on the experience of industries that were ripe for liquidation or semi-liquidation, and all the positive statistics that were cited were germane to that group. What we have heard here are two people talking about industries that are vital and need technology renewal, which is really the subject of this symposium.

Now, my thesis is that the short-term investment horizon is, in fact, the problem. There is, in fact, a cost of capital differential that is a problem, but I don't view that as the determining issue. It seems to me that the pernicious event is that the returns offered by liquidation and semi-liquidation—and the speakers were liberal in quoting them, minimum 30 percent, maybe 50 percent—are now substituted for what were the traditional 9 to 10 percent equity returns, and those are now being imposed upon companies that have a much more vital mission. A lot of the commentaries by the government people and the academic observers are confusing the two. You can't operate a vital institution on the kind of returns you can get out of liquidation and semi-liquidation.

MR. BOOTH: I tried to say the same thing. I think the 30 percent returns are there for private benefit. I think we all should recognize that that is our system. I mean, to ask a person who did an LBO, is it a good idea, is sort of silly to me because they wouldn't have done it if they didn't think it was a good idea. They think they are going to make a lot of personal money out of this thing: that is why they are doing it, and it is a short-term focus. I really do believe that.

So, I believe we are talking past each other. I would agree with you.

MR. PAXTON: I am Harry Paxton from Carnegie Mellon University.

Most of my career has been spent in research and I like to talk about it to anyone who will listen. So, I am very interested in your remarks, Mr. Booth, that you would like to tell a number of constituencies about research. My experience has been when I go to a cocktail party and someone says what do you do and I say I am in research, their eyes glaze over. Do you have any clever thoughts on how you are going to convince your constituencies?

MR. BOOTH: Well, I usually don't see my owners in the cocktail parties, but I didn't find that to be true personally when we went around and visited with our owners. We spent a lot of time with them. I think what Henry talked about is true; sometimes they are the owners of the day. I mean, there is a tremendous change in shareholder ownership when you are in this sort of a play that we were in. We went from 20 percent to 60 or 70 percent of people that were less than a week in their holdings.

My sense is that research isn't as important to them as what you are going to do next week and how you are going to recapitalize the company and pay them a dividend, but I do believe that there is a genuine interest in where your company is going and what your research is developing. I didn't find them glazing over. I found them very interested, but we had to spend a lot of time and a lot of effort and a lot of visits to our laboratories and a lot of Peter Kliem's time and a lot of our scientists' time to really take them through it. I found that a product of that was a positive response.

So, that is all I can tell you from personal experience.

MR. EIZENSTAT: Henry?

MR. WENDT: I would only say that the follow-up question usually is, what is coming out of research? And if you describe a product that is going to be introduced in the next quarter, the eyes brighten right up, but if you describe a wonderful breakthrough that might make it onto the market in 1994, they turn on their heel; if they are shareholders, they go right out the door. That is all part of this phenomenon we are talking about.

MR. ZAININGER: I am Karl Zaininger from Siemens. I would like to thank these panel members, first of all, for addressing the topic of this meeting. It was very nice. I didn't think that the first session did that.

Secondly, I agree with you on just about everything, but I would like to ask Mr. Wendt a question, which is an extension of his comments on globalization or transnational companies, and that is really the concern about the global management of R&D. You just talked about R&D, but the question is, as you become more global where do you do what research and why?

MR. WENDT: Well, that is an excellent question and if I really had the answer to it, I guess, I would have written a book or something, but the thrust is that the output has to match the market but the source of

R&D does not. So, one tends to put one's investment in a place where you expect continuing productivity of the science and technology.

Historically, the United States has been a very good place and I believe will remain a good place and I pointed to Britain also as a good place. Frankly, we are probing Japan, to be perfectly honest about it. Japan may be or may become a good place for the new non-traditional sciences, as opposed to the more traditional sciences. So, in biotechnology, for example, or molecular biology, particularly using fermentation technology, Japan is very interesting. They have great expertise in fermentation technology and they are applying it to this new technology.

In the more traditional technologies, frankly, I think Japan is a terrible place to do research. The academic hierarchy is very, very restrictive. It is very difficult, particularly for a foreign company, but in the new technologies, we are intrigued. I mentioned a collaboration with Suntory. That is kind of a scientific probe in that direction.

So, that is our view at the moment. The Continent has some appeal, although we can do in Britain much of what we can do on the Continent and probably obtain better value, because we have the investment there and the people there already. But it is a question we ask repeatedly and a large company, I think, tries to push the envelope on that basis.

MR. EIZENSTAT: May I ask a related question, Karl, just to follow up on that because this is, I think, a concern of people who are concerned with employment. Is it naturally the case in this new global market to both of you that when you move your R&D facilities or some of your R&D abroad that it will naturally bring manufacturing along with it? Or will you perhaps just do research in the best place possible and the manufacturing may be done in quite a different place?

MR. BOOTH: Let me just say we don't do any research outside the United States. We do do manufacturing, and in our case it would follow the other way, that the research would follow the manufacturing more than the other way around. We will be questioning that as time goes on, as to how much research we do want to do outside the States, but in our case it would be the reverse of that.

MR. WENDT: Well, it is a very pertinent question for SmithKline Beecham in that as a consequence of the merger of these two very large companies, we are examining everything that we do with an eye to the future rather than to the past. In the past, the practice in the pharmaceutical business has been to put a plant in almost every country, because the regulatory authorities of the country like it that way. They can regulate it. If the factory is not there, they feel they can't regulate it so well and so they are unhappy and they make life hard for you.

I think that is clearly changing in Europe. So, the answer now and for the future is, you do your research in the best place and you do the manufacturing in the best place. You manufacture in the fewest places possible, in order to gain the greatest efficiencies and the greatest competitive advantage.

MR. EIZENSTAT: And they may be two different places?

MR. WENDT: And they are almost certainly two different places. Scientists can get on an airplane, just like anybody else and go to the factory and vice-versa.

Going back to the other question, one other thought.

There is no large biologically-based corporation, health care corporation, pharmaceutical or otherwise, that can afford not to do research in the United States. That should give you all some comfort. Beecham was looking at a very major investment in the United States that this merger made unnecessary. It is absolutely competitively vital, at least in the health care business, to do research in this country and I think that most large Japanese competitors—most of them do not do any research in this country—if they were in this room and had a little *sake*, they would agree with that and say so.

MR. STERN: I am Bob Stern, an independent consultant. My question is largely addressed to Mr. Wendt, but really to both of you.

I was struck by somewhat negative connotations in your views about consortia as a way to do research and, yet, the Europeans have had a fair amount of success at it through things like Esprit and Eureka. The Japanese in their superconductivity consortium have had a fair amount of success. I wonder, in this globalization era, whether that isn't a form that the United States need to experiment with, and whether we need to gain experience before we look at it too negatively?

MR. WENDT: Well, I will start in response.

My comments were directed to the concept of an American consortium in competition to other countries. We do collaborative research in a variety of ways, as I think I indicated in my talk, and we will certainly continue to do so. While I can't think of a case at the moment, I could imagine that we would also engage in a consortium if it proved desirable.

I would definitely prefer to do that, however, with a global view and I would want my partners to be the very best in the world. Now, if the very best in the world are all American, that is great; I am all for it, but if the very best in the world are also from other countries, then that is the way we have to go. We have to do that to be truly competitive. To think that we can just sort of erect national boundaries in this day and age, I think is wrong.

MR. STERN: It may be that the economies of scale for R&D may drive us in that direction anyway. Something I have noted is that, for example, with regard to the semiconductor area, it is very important to U.S.-based corporations to be members of JESSI, which is the European counterpart. It seems to start out with a national goal in mind—certainly that was clear in the Japanese case—but I think maybe everyone will have to broaden their view and do what you say, which is get the best researchers in and look at the overall impact. Maybe that is the way it will go.

MR. WENDT: Whatever is best I think has to rule. I agree with what you said.

MR. EIZENSTAT: In fact, I think one of the arguments that has been used against opening Sematech to non-U.S. companies has been the fact that the European consortium has been closed to companies like IBM, which tried to get in.

Henry and Mac, there may also be a question about whether government-led consortia work better in places like Europe and Japan, which have a stronger, longer history of deeper government involvement in the economy; whereas, there is a certain distrust in this country of that. That is an open question that we could debate, I suppose, forever.

MR. ALTHUIS: I am Tom Althuis from Pfizer, Incorporated, and I would like to address a question to Mr. Wendt.

I was particularly fascinated by your thoughts on the transnational corporation. However, your company and mine have long been international or multinational corporations and I am not quite clear what you view as the differences between the transnational and multinational corporation—although I think some of that began to come out in the answer in the previous question—and what that specific difference is going to have on research.

MR. WENDT: Thank you. I am assisted in answering that question by the fact that transnational appears in both *Webster's* and the *Oxford English Dictionary* and the definition is "exceeds national boundaries." Multinational, also, in both of those dictionaries is defined as operating in more than several nations. So, in brief, that is the difference.

Now, SmithKline Beecham, more than SmithKline or Beecham prior to the merger—and I dare say, perhaps more even than Pfizer—has sales and profits that on a proportional basis match more closely the global market. I mentioned that Europe is the larger market, followed by the United States and Japan.

Basically, our sales and profits match the global market and are not dominated by any one country, although the U.S. is the largest country, Europe is the largest market. Our ownership is 50/50 across the Atlantic Ocean; therefore, not dominated by any one country. Our board is 50/50 in terms of national identity; therefore, not dominated by any one country; likewise, the management. Our purpose is to bring onto the board more third-party nationals, to truly reflect at the highest levels of the company this transnational character.

Our R&D is split almost 50/50 between the United States and Britain in the new company; therefore, not dominated by any one country. I think that those various factors go pretty far in illustrating the dictionary definition of "exceeding national boundaries." I don't know that I can say that it is a prototype for the future, but it is the way I think the world is going and this company illustrates it. There are a few others that are moving in that direction certainly.

MS. BLAIR: I am Margaret Blair with the Brookings Institution.

I want to go back to the issue that both of you have spoken to, about the short-term mentality in the financial markets. What I want to get at is, I want to relate that back to something that both Mr. Altman and Mr. Grundfest talked about last night: the high cost of capital. My own research has been looking at the possibility that it is the high cost of capital that is in fact driving the restructuring, because the high cost of capital enforces a short-term mentality on the investor. The question I want to ask is about which way the causality runs, because it makes a big difference in terms of policy implications if you think that you have got a short-term mentality in the stock markets, and that is what is driving up your cost of capital, or if there is some exogenous factor driving up your cost of capital and that is forcing people to think short term.

Let me tell a little story to illustrate my point. I recently had an opportunity to look at some research that was done by a Japanese economist, who is studying at the Brookings Institution. He decided to see if he could come up with some explanation for why the performance of Japanese manufacturing companies had been so far superior to American manufacturing companies during the last 15 years. His measure of this was that the Japanese have gained market share and the American companies had lost market share.

He went through several different explanations for why this might be and concluded that it wasn't that the Japanese companies had a lower materials cost and it wasn't because they had a lower labor cost and it wasn't because they had some sort of exchange rate advantage. It was simply due to the fact that the Japanese firm had a much lower cost of capital. When I looked at his results, I said a Wall Street economist would look at that and say, in fact, American companies perform much better than Japanese companies because their return to capital was higher during the same period.

It seems to me that it makes a big difference in what your goals are and what questions you are asking here about what you want out of the company. So, I think either or both of you could address this question of which direction the causality runs. Are we thinking short term and, therefore, demanding that our firms impose a higher hurdle rate on our their investment strategies, or is there some exogenous factor that we could be addressing that is driving up the interest rates and, therefore, forcing us all to think short term?

MR. BOOTH: Well, I am not sure which is the chicken and which is the egg. I am not sure which way that goes. I think early this morning there was just a slight mention made of the fact that many of us are invested in pension funds, which demand high returns. They are a huge source of funds in this activity that is going on out there. It is kind of like eating your own tail to a degree here. I mean, if we weren't demanding the 30 to 25 or whatever it is percent return on our investment in our pension funds—I think that is where the cycle starts.

Now, does it start there because someone can deliver that and, therefore, it keeps going? I am not sure where that starts, but I do know that it is a huge source of the funds that are in all of these big, highly leveraged situations. I think this idea of making folks pay more dearly for short-term transactions is a very good idea. I think that may change the mentality. I think we are on a roll here that we don't know how to stop.

The financial people—I have met with many of them, as I am sure Henry has, and if they don't get us 30 percent, they are out. The pension fund cuts them off and they are off to someone else. So, that creates a very barracuda kind of a thought process.

MR. WENDT: I really don't have anything to add to Mac's reply. I don't know which is the causative factor either. I hope you find out, though, and tell the world because the phenomenon is very dangerous. It may, indeed, be caused by us and our pension funds. And there, incidently, total tax exemption is also a factor. They just roll their money as rapidly as they can.

MR. HILL: I am Chris Hill with the Congressional Research Service. A question for Mr. Wendt.

I am glad you introduced the analysis of the transnational corporation into this morning's discussion. My suspicion is that that phenomenon is a great deal more important in the long run to research and technology development than the LBO phenomenon, per se, is likely to be. And let me say, I agree with the substance of your analysis of where we are and where we are going.

But I need your help. I work for Congress. Congress needs your help in the following little conundrum. We are now spending about 20 to 22 billion dollars a year in public money on non-defense-related research and development, about 6 billion of which goes for purposes that lie at the root of your own industry, the biomedical research at NIH and elsewhere.

A lot of the argument about why that is a good investment of United States citizen-taxed public funds, or borrowed public funds, is that it will enhance the competitiveness of American firms in the competition with firms abroad, and contribute to the income and standard of living of the working people of America and so on.

We need some new arguments to be able to say why that continues to be a good idea in a world of transnational corporations. Or are we, in fact, on the verge of having to reexamine the relationship of the transnational corporation to the independent nation-states? Do we need new parallel structures of governance, so that there continues to be some overlap between the geographic domains of industrial operation and those of taxing, spending, employment and unemployment?

MR. WENDT: Well, I definitely endorse the latter view in terms of advice. I think that you would also be well-advised to reexamine the tax policy with respect to the deductibility of R&D against foreign-sourced earnings, as opposed to just U.S.-sourced earnings, which for any globally-minded company tends to drive research right outside the country. It seems to me to be totally counterproductive, and a good place to start.

The American pharmaceutical industry is now spending more on research and development than you in Congress are spending at the NIH. Those lines have now crossed, with NIH really plateaued—in real terms, I think, down—and the industry up. Nevertheless, I think we would all say, and certainly our R&D executives would say, that NIH has served as a wellspring of scientific excellence and training for the country.

Like all big institutions maybe NIH, too, should come under the discipline of restructuring; it is fair to say that it hasn't. So, if the thrust is on training and developing first-class scientists and practicing leading edge science, I think it is a good thing. That proposition really has to be examined with respect to our national-funded research centers. That is about all the wisdom I have and it is not very much, I am afraid.

DR. MUELLER: Just a brief comment. I am Dennis Mueller at the University of Maryland and also will talk this afternoon, but I will save myself having to make a detour this afternoon.

Just following up Margaret Blair's question, because I think it is a very important one. My work and work I have done in collaboration with the Japanese does suggest that both U.S. managers and Japanese managers are pursing growth, perhaps more than would be in the wealth-maximizing interests of stockholders, as has been mentioned earlier this morning. The important difference is, however, that very often U.S. firms have done that via the merger route and via the "take over all other companies" route, far more than the Japanese. Indeed, we have 10 times as many mergers as the Japanese do. They have relied much less on mergers and acquisitions as a growth strategy; rather, they have relied on investment R&D and taking over markets. And of course, they have been very successful at it.

Thinking about policies and how we change things around in this country, the thing to be asking ourselves is how can we get managers to shift emphasis from taking over other firms to taking over markets, and how to get the growth proclivities of management in this country directed in what is more socially beneficial investment in R&D kinds of investments.

MR. WENDT: Is that a question?

DR. MUELLER: If you have an answer, it is a question.

MR. WENDT: I would start the examination with analyzing the difference in the shareholder structure in the two countries and the demands of shareholders on the management. And if anybody thinks that corporate managers are immune from shareholder demands in 1989, they are crazy. I think that is where it starts frankly.

Mac should give his perspective, but that is certainly mine. The structure is obviously very different in Japan, the needs and demands of the shareholders are very different, and it is no wonder that there is a difference in managerial behavior. That is who we work for, that is who the directors remind us that we work for, at least once a month; generally, more frequently.

MR. BOOTH: I absolutely agree with that, but there was something else that you were saying I would like to comment on. It has to do with the perception of what happens in Japan versus the United States as far as collaborative research. You don't need mergers and acquisitions if you can get collaborative research. I am not an expert on the Japanese marketplace or Japanese industry. I can speak about my own experience and our own corporate experience in electronics and in photography.

There is a tremendous amount of collaboration that everybody denies, but it goes on continually in the marketplace, about sharing technology. It is one of the most remarkable things I think I have ever seen, where they kill each other in the marketplace and yet share technology between companies and have common suppliers that both invest in to provide them with components. It is a remarkable arrangement that we haven't done. Our antitrust activity and all seems to squelch it historically, and I don't think it is probably in our culture, but they do it.

So, when you say they don't do mergers and acquisitions to do that, I am not so sure—I am sure that is correct literally, but if you look behind that, I am not sure that the same things aren't going on.

MR. WENDT: Just to tie those two responses together, I think Mac is describing the *keiretsu* system of collaboration, often apparently informal, but driven essentially by common shareholders. That is what provides that. That is the umbrella for that system and if they are not common shareholders, they tend not to cooperate in quite that fashion.

MR. EIZENSTAT: And they are often big banks who don't demand the short-term returns.

I really want to thank both of you. This was a terrific contribution to our topic with very rich discussion. We appreciate your coming.

# Afternoon Session
# October 12, 1989

MR. EIZENSTAT: Let me introduce Ken Flamm. Ken, as many of you know, is Senior Fellow in Foreign Policy Studies at Brookings. He received his A.B. in economics from Stanford and his Ph.D. from MIT. His most recent research has been concerned with the determinants of international trade and investment patterns in high-tech products. He has written several books in this area: *The Global Factory*, *Targeting the Computer*, and *Creating the Computer*.

Most recently he contributed a comparative study of the rate and determinants of technological change in the computer and communications industries to a Brookings publication entitled *Changing the Rules*, a conference volume co-edited with Robert Crandall. It examines the impacts of technological change, internationalization, and deregulation on the structure of the computer and communications industries.

Ken is a recognized expert on technology policy in the United States and is also the author of the background paper that was distributed prior to the symposium, which gave us a very good context for our discussion.

MR. FLAMM: Thank you very much. I basically want to talk about four things today, all of which relate to the subject of our discussions this morning. First of all, I prepared the background paper you received, and that was an attempt to summarize the existing studies on the subject that has brought us here today. I would like to talk about some evidence on a subject that was not talked about in that paper—a very brief, crude pass at some evidence.

Secondly, I would like to mention briefly the overlooked factor in all of our discussions here today, and that is the Federal Government itself, which has an impact on research and development in this country through its policies. It is not just Wall Street that shapes and alters the allocation of resources to research and development in this country, and I would like to talk very briefly about that.

Thirdly, I would like to point out some of the major issues that came out of various studies on the impact of takeover activity on research and development, pointing particularly to a couple of subtleties that are not always evident to the naked eye.

Finally, I would like to make some brief remarks on capital costs, a subject that has come up over and over again, as linked to this entire discussion. Although I am no expert on capital costs, as a hobby I try to collect studies comparing capital costs in the U.S. and Japan. Let me just sort of dump on you my casual perspective on those, because I think some of that discussion is relevant to issues that have come up today.

Without further ado, let me start in on the "new evidence" aspects. In discussions of research and development expenditure, those who argue that takeover activity has had some effect on research and development activity in general—that is, a shift toward shorter-term investment projects—also almost inevitably talk about a shift within the category of research and development. The shift, they say, is away from longer-term projects, in particular, basic and applied research, toward projects with shorter-term pay-offs, that is, development expenditure.

There is very little evidence on this particular point; in fact, I am unfamiliar with any. What I would like to do with you today is to briefly share with you a graphic I made up that reflects some quasi-National Science Foundation data. What I have done here is, through I hope artful calculation, filled some empty cells in the NSF data to put together a time series.

[Editor's note: Mr. Flamm's graphics were rough sketches prepared during his talk. They were not recreated for this publication because Mr. Flamm's explanations of the data are clear without them.]

This graphic basically looks at company-funded research and development in various U.S. industries and attempts to calculate what percentage of that is going to research as opposed to development. If there indeed has been a shift in focus away from longer-term projects and/or shorter-term projects, the share of research declines. Unfortunately, and this is the first of a long string of "unfortunatelies," the National Science Foundation data really only go up to a relatively distant period. It is about 1985, before much of the activity we are talking about hit its peak. Nonetheless, I think it is worthwhile to point out a few salient facts.

The bold line is the average for all industries. These data are also broken out by industry, in terms of research share—that is, companies' funds for research as a share of all funds for research and development. What I would like to point out is that overall, for all industry, the trend has actually been pretty flat but for particular industries there has, in fact, been some evidence of slight decline. However, it seems to have preceded the really dramatic increase in takeover activity in 1984, 1985 and 1986. Overall declines in the machinery industry, in the chemical and refining industry, pharmaceuticals, and instruments were offset by increases in the share of research in company R&D funds in the aircraft and missiles industry—aerospace—and in electrical equipment other than communications equipment.

I would point out in passing that those are both sectors that enjoy heavy government procurement. There has been a big increase in defense spending, so one interesting hypothesis that perhaps goes with the observation that government procurement can be a driver for company research—as Frank Lichtenberg in particular has pointed out in his work—is that the rise of defense spending around this period may have played some role in the increase in research in those industries. But to make a long story short, there is not a great deal of evidence to suggest a steep decline associated with the rise in takeover activity, although obviously these data are very limited.

The second thing I wanted to point out is that when we are talking about how the Federal Government can influence research and development activity in this country, clearly one avenue is policy aimed at regulating or restricting takeover activity. That is certainly the focus of much discussion today, but I would just like to point out that the Federal Government funds a substantial amount of research and development activity in industry. In the process of going through some recent figures, I was struck by the relatively large percentage of research and development resources in industry that are absorbed in government-related work.

That, of course, immediately raises the issue, is there some way to make this substantial federal investment in research and development more relevant to the competitiveness concerns we have been talking about? I would just like to note that overall, something like 18 percent of basic research carried out within industry is funded by the Federal Government. Of course, there is a substantial variation across industries with aircraft and missiles and electrical equipment showing relatively high percentages. Non-manufacturing, somewhat surprisingly to me, is extremely high. I am not sure what is in there exactly, but this is a pattern that emerges in many of these statistics.

Just for the record, here is applied research: similarly relatively high percentages, particularly in the machine industry, and the computer industry obviously accounts for much of machinery. And, finally, even in development you find a substantial amount of resources being absorbed in work for the Federal Government. The overall average is about 36 percent, ranging from about 3 percent to 78 percent. Again, federal funds in the chemical and pharmaceutical industries are not as notable; the Federal Government plays a relatively small role, as opposed to aerospace, electrical machinery and other industries. "A lot of variation across industry" is a message that perhaps ought to be stressed more in today's discussion.

That is it for the viewgraphs except one. The third thing I wanted to do today is to point out some of the subtleties in the empirical evidence on the impact of takeovers on research and development activity, which may have been glossed over to some extent in some of the discussion we have heard thus far. Without having to actually do this research, it is easy to read and criticize other people's studies and point out shortcomings. But if you look at the arguments about takeovers, it is not clear that all merger and acquisition activity is created equal. Certainly, some of the arguments by opponents of leveraged takeover activity, in particular, hinge on the impact within a company of the increase in leverage on research and development activity.

By and large, the empirical studies that exist do not attempt to distinguish between firms that greatly increase their leverage as a result of takeover activity and those that do not. Yet this, it seems to me, is a variable you would want to look at if you were interested in that particular proposition. So I would argue there is a considerable amount of research that remains to be done if these things are to be nailed down quite firmly.

A similar comment applies to some of the work associated with the views of Michael Jensen and his colleagues, who argue that hostile takeovers in particular are a vehicle for reorienting management to the stockholder interest. Obviously, friendly takeovers—in particular, management buyouts—ought to be regarded as a somewhat different animal than a hostile takeover, if the intent of the activity is not to reorient or replace existing management. If one wanted to look at the impact of hostile takeover activity on research and development, one would want to distinguish between hostile takeovers and non-hostile takeovers. Again, it is not clear that all merger and acquisition activity is equal if you are looking at the arguments which lie behind the assertion that takeovers are either good or bad.

The second point that I think is subtle, but well worth making, is that by and large the data show that there is no strong evidence that research and development intensity declines as a result of takeover activity—that is, that the ratio of research and development expenditure to sales declines. In fact, there is even some evidence that it may increase.

However, it is not at all clear that this is the same as saying that takeovers have no effect on research and development. Why? Because it is entirely possible that you could have stable or even rising research and development intensity, yet have an aggregate volume of research and development that declines, in particular if the size of the firm declines. Now, since much of the literature on takeover activities suggests downsizing of the firm in the wake of a takeover, it therefore seems clear that it is not quite right to equate research intensity with the aggregate size of research and development activity.

Some of the confusion, I think, about empirical evidence on the subject comes from making statements about research based on one variable or the other when really they can behave quite differently. In particular, I would argue that it is quite likely that some of the apparent discrepancies between the studies produced by the National Science Foundation and the studies produced by academic researchers hinge on this distinction between research intensity and the aggregate level of R&D.

This point is particularly important because both the schools of thought on the subject essentially argue that some decline in research and development activity is likely. Those who believe that takeovers are a positive development argue that unprofitable projects are being eliminated and that some of those projects undoubtedly are research projects, and that therefore you would expect research to decline. Those who believe that increased leverage causes declines in research and development activity and who view this as a negative development for the United States, also argue that takeovers cause declines in the total volume of research and development. Both sides seemingly concur on the proposition that takeovers should cause declines in the overall level of research.

Well, that, of course, raises an interesting question. If this is true, why is it that most of the empirical studies seem to show no such relationship. The answer, as I pointed out a moment ago, is that most existing empirical studies look at research intensity, not the overall level of research and development. These two seemingly intractable and contradictory propositions can in fact be reconciled.

A third subtlety which is important, very difficult to grapple with, and in some sense the crux of the matter, is that those who argue that takeover activity has basically served a positive function, argue that much of the positive function has been exercised through what you might call demonstration effects. That is, even if you are not subject to an actual takeover attempt, you will look at your colleague at the company next door who was, and you get to thinking, maybe I ought to be thinking a little bit harder about what I am up to.

In all of the empirical studies which I reviewed in the background paper prepared for you, the methodology basically boils down to comparing firms that are merged or taken over with firms that are not merged, not taken over, with different definitions of merger and takeover. But in comparing those two groups, you get no handle on whether demonstration effects may have been important. That is, both groups may have declined and there may be no obvious difference between the two groups that is evident after the decline. Yet it still somehow may be a product of buyout and takeover activity. This is another subtle point that I think is very difficult to deal with; as I understand it, Bronwyn Hall is in the process of working on a study that will deal with it. I look forward very much to hearing the results of her investigations on the whole issue of demonstration effects.

The final point I would like to make today relates to capital costs. I would like to reiterate the kinds of connections that Margaret Blair was making earlier in her questions today. Along with her colleague and mine, Bob Litan, at Brookings, she has been making what I regard as quite a plausible argument. I would like to briefly sketch out that argument in a way that perhaps is a little differently stated than you have seen thus far. I think it can be best visualized in a diagram. The diagram looks something like this. Let us imagine for a moment that real capital costs have risen quite sharply in this decade. Of course, it is not going to take a lot of imagination to assume that; real capital costs have gone up. I am not really doing justice to the first part of this argument. I urge you all to consult a very interesting paper done by my colleagues Blair and Litan at Brookings, which spells it out in much greater and much more persuasive detail than I possibly could.

The bottom line is this. As real capital costs go up, what should the impact be on a rational businessman? The answer is, longer-term projects like research would suffer, presumably because the hurdle rate would go up and they would be much more affected than shorter-term projects. Therefore, in general we expect long-term projects of various sorts, and in businesses that were being developed with long-term expectations in mind, to decline. Presumably research would be especially impacted—particularly more basic kinds of research—because of the long-term nature of that kind of activity.

According to this scenario—and here is where it ties in with Michael Jensen's arguments on the salutary effects of takeover activity—if we are coming from a world of relatively low capital costs, into a world of relatively high capital costs, one would expect pressures on companies to adjust, that is, to cut back on the longer-term projects and to divert more of their attention to shorter-term projects.

In this world of higher capital costs, some companies perhaps would take longer than others to adjust. As a result, you would then expect pressure on the part of the stockholders to speed this adjustment to reflect their interests in earning a higher return on their investment, in this world of higher real capital costs. As a result you would begin to have hostile takeover activity, conducted with the objective of liberating cash flow from those firms, to be invested by stockholders in more profitable activities than perhaps are available in the industry—or perhaps just buying Treasury bonds (federal borrowing may be one reason why capital costs have gone up).

So hostile takeovers in turn would have a demonstration effect. The example of firms actually subject to hostile takeover activity would in turn have impacts on other not yet threatened firms, and again lead to a process, through demonstration, of cutting back long-term projects elsewhere in the industry.

So if we are going to ask the question, what is causing the decline in research in this scenario, there are a couple of different ways you can answer the question, all of which would be correct. You can say hostile takeovers are causing a decline in research; strictly speaking, you would be correct, because they were the vehicle by which demonstration effects influenced everyone else in industry and got them to cut back on research and development. Or you can say that real capital cost, the macroeconomic environment, was the culprit, because it was causing this diversion away from long-term projects toward shorter end projects, and you would also be right. It is just a question of the chain of causality and where you choose to break into it.

Also, let me just reiterate that discovering that research and development intensity does not decline is completely consistent with aggregate research falling. So this story could indeed be true even if nothing at all has happened to R&D intensity.

The final point which I think merits at least some brief discussion, because it is very much central to many things that have been said today, is the discussion of capital costs and their impact on research and development. In particular, the differential in capital costs between the United States and Japan, which is often pointed to as one of the major structural factors behind our relative decline vis-à-vis Japan, in industrial success.

I guess the major point I would like to make is that there are many, many studies that have been done thus far on the cost differentials for capital between the United States and Japan, and most—but not all, I should add—of these studies conclude that capital costs, the real costs of funds and the real cost of capital, in Japan are substantially lower than they are in the United States.

The evidence on the margin of that difference is actually quite mixed. In fact, last night you heard Joe Grundfest refer to a 100 percent differential. Talk is cheap in Washington, and I have attended various meetings where I hear people mentioning 300 percent, 400 percent differentials in the cost of the capital between the U.S. and Japan. I would add that there are other studies that point to substantially smaller differentials, on the order perhaps of 20 to 25 percent. The reason you get such widely varying numbers is that there are such difficult methodological issues involved in comparing the costs of capital in the U.S. and Japan. Many of the reasons were brought up earlier, having to do with the different structure of shareholding in the United States and Japan, imperfections in the Japanese stock market in particular: extensive cross holdings, segmentation of the market, and various other things.

Most studies that I know of conclude that the cost of debt is roughly similar. There is some difference but it is relatively minor between the U.S. and Japan. Intuitively we know this has to be true because, according to my colleague Ed Lincoln's figures—he is our Japan expert at Brookings—something like half of all the bonds issued by Japanese corporations were floated in international capital markets. It really does not make much sense to borrow abroad if capital, or at least debt issue, is that much cheaper in Japan.

So if we assume that the cost of debt is at least in the same ballpark, that means that any differences in the cost of capital are going to boil down to differences in the cost of equity. There we have very difficult issues, particularly if you think about the problems of measuring the cost of equity when you have highly imperfect or segmented markets for equity.

If we assume for a moment that there are enormous differences in the cost of equity in Japan, then it seems clear that the rational response would be for Western companies to issue stock in Japan and tap into those very cheap equity markets in Japan. Obviously for this not to happen, there must be structural barriers or imperfections of some kind. These very large differentials in the cost of capital simply are inconsistent with perfect financial markets. So one way or another you have to end up saying something about imperfections in the Japanese capital market; that greatly confuses the issue and creates a cloud of uncertainty over estimates of the relative costs of capital in the United States and Japan.

That basically wraps up what I had to say. If you took the trouble to read through my rather circuitous review of the empirical evidence that has been adduced about the issues we are examining today, you can see that I spent some time puzzling over the various inconsistencies and trying to square different pieces of evidence that seem inconsistent. I have argued today that the empirical record is far less certain than some would have you believe, in part due to subtle effects that are not really measured in these studies, and more important, because of the distinction between research intensity and the aggregate level of research.

MR. EIZENSTAT: Thank you. I must say that in hearing the economic viewpoint, it takes me back to my White House days where we wanted to have answers and we often got questions on both sides, because the data are difficult to get and that is the nature of what we are dealing with.

It really is an enormous personal pleasure and privilege for me to introduce Dick Gephardt to you. I have known him since his first term in Congress, and his truly meteoric rise is nothing short of spectacular. I think one would have to go back an awful long way in American political history, to the very beginning to find anyone who, in his seventh term, had been elected to the second most powerful position in the House of Representatives.

He was first elected in 1976 and because of the respect of his colleagues, quickly rose in House ranks. He was Chairman of the House Democratic Caucus, which was a particularly challenging position during the Reagan Administration, and it really put him at the forefront of shaping the Democratic Party's platform. Dick found time in the midst of pulling all of those policies together to put his stamp on what is perhaps the most important tax bill of the last 20 or 30 years, the 1986 Tax Reform Act, which is properly called the Gephardt-Bradley Tax Bill.

What happens in the next few weeks will determine, in a sense, the longevity of that effort, and, of course, that is one of the reasons I think that Dick has been so much up front in talking about some of the tax issues, capital gains and the like, that are in front of the Congress now. When Dick ran for President in 1988, to me it was the best example of the respect of his colleagues because at least a score of them took almost unbelievable amounts of their personal time away from their own campaigns to campaign around the country for him. This was more, I can assure you, than just a token gesture to someone who might be elected President. It was a genuine show of affection and respect and I still believe that a few votes here or there on Super Tuesday would have propelled Dick, after his showing in Iowa, to the nomination.

The *Almanac of American Politics* has said, and I think rightly so, that his major assets are his rare ability to come up with original and defensible solutions to complex and seemingly insoluble problems, his openness and his intellectual honesty. That is why we are particularly pleased that the Majority Leader has agreed to join us today to tell us how his colleagues view the wave of corporate restructuring that we have been talking about, and what we need to do in a broader sense to ensure U.S. competitiveness and particularly our commitment to R&D. We have here a leader in this country, and a person who will be a leader for decades to come.

Dick, thank you for gracing us.

CONGRESSMAN GEPHARDT: Thank you very much. I am very happy to have this opportunity to be with you this afternoon to talk about this very important subject. However, I do not want to begin without paying my respects to Stu Eizenstat. He said that I came in 1976. You will all remember that he did as well, and the fondest memories that I have of the Carter presidency were the times that I had to talk with him, to work with him, to think with him, even though we did not often get much time to do that, about what the policies would be and what we would try to do, and my admiration and respect for him have only grown through the years as I have continued to work with him on many different subjects.

I also want you to know, as I start this talk today, that when you have run for President of the United States and failed, you have a difficult time thereafter with so many people in the country knowing who you are, or thinking they know who you are, but not being quite sure. I will just tell two quick anecdotes that I think sum it up for you. A fellow in an airport the other day, I think it was Atlanta, was walking along very quickly beside me, and I was late for a plane and walking fast and he kept staring at me and then looking ahead and staring and I knew it was one of these deals where he kind of knew but did not know, and finally he just blurted it out, he said, you are Jack Kemp, aren't you?

The best of all was the other day in Florida. I was down there for our congressional candidate for Claude Pepper's seat, and I was coming down in an elevator in the Doral Hotel. Some tourists got on. They had bathing suits on, and they kept looking at me and I knew we were into one of these deals again. We got off the elevator, they took one of my aides to the side and were talking to him, and he came over to me afterwards and he was laughing wildly and I said, what did they say? He said, they said, you know, Vice President Quayle looks in person just like he does on TV. So I am not sure that the run for the presidency was worth the time.

President Kennedy used to love to tell the story of the man who was seen crawling around on his hands and knees under a lightpost in a parking lot. When a passerby came by and asked him what he was doing, he said, I am trying to find my car keys and it is dark. The passerby asked the man if he could recall where he had lost them or where he had had them last. He said, oh, that is easy, I dropped them just as I was getting them out to open my car door, as he pointed to his car which was on the other side of the parking lot. So the passerby said, why are you looking over here? The man said, because the light is better over here.

Many of us have become convinced that America has lost something very important, its economic preeminence in the world economy. And yet, for all of the 1980s, we have insisted on looking not where the answers truly lie, in my view, but where the light is better. I salute all of you for having the courage to look where we need to look, not where it is politically convenient to look.

For the last several years, I have been fighting to elevate our trade and economic problems to the top of our policy agenda. During this battle, I have come to realize that one of the principal problems is our unwillingness or our inability to think about our economic problems in more than one dimension. Everyone seems to want to find the silver bullet, the magic solution, the one answer that will restore our economic strength by itself and overnight.

Rather, I think what we need is a comprehensive approach that looks at the long-term needs of our economy and our country. A recent poll indicates that 60 percent of the American people see our competition problem being caused by too much focus on short-term profits, and more than 50 percent rate an excess of corporate takeovers as a major factor. Last year's trade bill that we passed, which has gotten so much notice for Super 301 and a lot of other very, very emotional topics, addressed many of the concerns facing our nation.

Your symposium addresses others, especially the troubling decline in industry research and development, the increase in corporate debt through mergers, acquisitions, leveraged buyouts. It is these and other long-term problems that must be addressed if we are to restore America's economic strength.

Of late, there has been concern that the wave of corporate restructuring has reduced the rate of growth of corporate research and development. I know the studies on this issue are inconclusive in the narrow scientific sense, and I am certainly no expert on this subject, but I know one thing. I know what I would do if I were a chief executive of a company under severe financial market constraints to show improved financial results on a quarter-to-quarter basis. I would be under pressure, extremely high pressure, perhaps unbearable pressure, to reduce all expenses—as well as research—which were not tightly focused on quickly developing profitable products.

Thus, at the very least, I expect the character of research coming out of industry to shift away from free-wheeling basic research to highly focused research. A climate like this might have made it impossible for Bell Labs, for instance, to discover the transistor.

So the question is, what can we do about it or what should we do about it? We cannot and should not construct barriers to all takeovers in corporations, but we should examine whether we can remove those tax subsidies that exist that artificially encourage takeover activity.

Second, the trend toward takeovers and leveraged buyouts and reduced emphasis on basic research supported by industry puts an even higher importance on increasing the level of federal support for basic research. The Congress has been trying to do this and will continue to try. For example, I believe that substantial increases in the National Science Foundation budget are desirable, and even necessary, even in the world of tight budgets. However, because of the budget problems we face, we have been unable to do what needs to be done, in my view, and will continue to fall short of our goals as long as those budget problems remain.

You also know that the deficit not only makes it difficult to increase federal support for research, but by keeping interest rates and the costs of capital high, the deficit makes it almost impossible for companies to afford the longer view that we all know is necessary. In my view, we will not solve our budget problem until we get presidential leadership on making the tough budget choices. Congress has a role, that is clear, an important one, but we only have one leader in America, and that leader must lead in order to get the budget to be changed.

Many are aware of President Bush's black box budget earlier this year [1989], where he specified a set of increases to the Reagan budget and let Congress do a multiple choice test on where to make the cuts needed to make it all add up. He said, read my lips, no new taxes, and while that clearly is the platform that he ran on and what he obviously believes, that also constrains our ability to put together the kind of commitments to investment in research that we need.

Third, I would like to spend some time telling you about the trade and technology initiative that I recently announced with Senator John Glenn, Congressman Mel Levine and Congressman Sandy Levin of Michigan, which I think will help American companies make better use of U.S. research resources by reshaping our trade and economic priorities. America's national security must be viewed not only in terms of defense capabilities, but also in terms of our economic prowess. The stark reality, we must realize, is that true strength today is measured in megabytes as well as megatons. Trade and technology will be as important to America's national security in the twenty-first century as air power or armor were in the twentieth.

Already we are seeing the proof. At a time when the Polish people are calling for our help to foster economic freedom and democracy, the President finds the U.S. simply does not have all the resources needed to lend a helping hand. Clearly, here and in many other areas we could talk about, economic and military security are inextricably intertwined. The trade and technology initiative will place our top economic officers on the National Security Council, ensuring that our economic security has an equal voice when generals and diplomats discuss national security with the President.

The bill establishes a permanent statutory position for the Assistant to the President for Science and Technology. Our bill restructures trade-related federal agencies, including reforming the Commerce Department into a streamlined Department of Industry and Technology, beefing up the foreign commercial service and creating an export strike force.

But what I want to discuss with you this afternoon is the part of our bill that deals directly with research and development, the part that I think is the most promising aspect of this new bill. As you know, the fundamental philosophy of federal involvement in R&D has not really changed much since the 1945 report issued by the Office of Scientific Research and Development which, in the heat of war, focused our priorities on basic scientific research and concluded that commercial applications were an area in which government should keep a strict hands-off policy.

Forty-four years later, I think that report is still right or to be more precise, about half right. Yes, the Federal Government must take the lead on basic research. No other entity has the resources and the long view needed to invest in it properly. But the other conclusion of 1945, that government should never fund or support anything with commercial applications, sounds to me as outdated as a 1945 fighter plane going up against an F-14, F-15 or F-16.

The good news is that throughout the 1980s, the federal involvement in R&D has changed. The bad news is it has changed for the worse. In 1980, the division between federal dollars for military research versus civilian research was about dollar for dollar, about where it had been for the previous 20 years. But by 1988, we were spending two dollars on military research for every dollar we spent on civilian research. Military R&D rose an astonishing 83 percent in real dollars during this decade and about 90 percent of the overall increase in federal R&D went to developing weapons systems. Non-defense R&D, on the other hand, declined in that period by 12 percent in real terms.

If we could return to the previous, albeit imperfect, ratio, if we could just have an even split between military and civilian R&D, it would free up an extra $15 billion. That is not spending more. That is just spending smarter. There are things we can learn from the military research program. One of the most interesting lessons we can take from it is structural. The Defense Advanced Research Projects Agency, DARPA, has done a wonderful job with a relatively small budget, which they use to perform high-risk, long-term R&D for the Pentagon.

Based on the DARPA model, we are recommending in our new bill the creation of an advanced civilian technology agency—a new way to promote the public-private partnership we need to compete on cutting-edge technologies such as biotechnology, advanced materials science and robotics.

ACTA, which will be its name, will have two aims. First, to assist and promote the development of advanced civilian technological capabilities and second, to assist in the application and commercialization process. That is, to help industries where they need it in getting new ideas from basic research to the marketplace. ACTA's focus will be generic research projects—areas of research, development and application that are not already being adequately addressed by the private sector alone and that are common to an industry or industries. These projects will be funded on a cost-sharing basis with industry. If the project makes money, each partner, government as well as industry, will get a share of it.

ACTA will be a lean organization, probably not more than about 35 professionals to begin with. They will assist in funding generic research projects where industry thinks there may be a potential for profit and a need for federal investment. It will not be directing or second guessing the market any more than the National Science Foundation does today. ACTA will be working with experts in the private sector on projects they consider to be promising. All market-based decisions will be left to those who can make them best, the private sector. In ACTA, as in all good partnerships, each partner will do what it does best. The government will put up the money, assistance, support, and leave the market decisions to the private sector.

ACTA could fund and support the technology transfers in joint ventures now being advanced by the machine tool, biotechnology, superconductivity and advanced television industries. It will encourage public-private partnerships through various levels of government, business, and academia, and significantly, ACTA will eliminate the sort of ad hoc, special interest funding on an industry-by-industry basis which characterized the birth of SEMATECH.

The key to ACTA's success will be to keep it as lean and as flexible as possible. I foresee people like yourselves taking an active role in places such as the Board of Directors and on various ACTA projects. You know where the action is, or where it should be in the next 20 years, but for various reasons, a given company or companies together cannot or will not go it alone. That is where ACTA will have to come in to help.

But this new idea is only a small part of the puzzle. We have a lot of do. We need to boost our national savings rate, improve our educational system, return a long-term perspective to both business and government, make it easier for industry to cooperate on R&D projects where desirable, and much more. We need to do all of these things yesterday. They are not simple, but neither are they impossible.

Our parents were able to defeat depression at home and dictatorships abroad at the same time, and while the fight is less dramatic today, the stakes are just as high. We have got to rise to meet the challenges of a changing world in order to give our children and their children the jobs, the hope and the opportunity they deserve. Einstein once said that splitting the atom changed everything except the way we think. So, too, does the new economic competition change everything. But too often, it does not change the way we think.

Last week, on Thursday I went to Europe, to Geneva, and then to Vienna and then to Brussels to be brought up to date, with the Arms Control Observer Group in the House of Representatives, on the status of arms control right now. I knew a lot about it, or I thought I did before I went there, but I came back with a much deeper understanding of what is happening in that important part of the world. It has been the scene of the two major world wars of this century, and I came to understand in a much more impressive way what the possibilities were.

For instance, I had not understood—maybe I had heard it but I do not think I understood it—that the Soviets were actually making an offer in Vienna in conventional talks that would reduce their tanks from 60,000 to 20,000 in one fell swoop; that they were offering to cut the number of troops they have in Europe and to demobilize them from 600,000 to 300,000. Also, to get rid of lots and lots and lots of other weapons. One of our NATO ambassadors, during one of the dinners, came over to me and he said, "Do you people in the United States really understand what is happening? It is a watershed. It is the end of an era and the beginning of a new era." The Soviets, I think quite clearly, at least under this leadership, are willing to make an almost irrevocable decision that the Cold War is over, the military conflict as we have known it has ended, and that we are now on to a new competition that will be economic and not military.

I then visited our people at the European Community and we talked trade. As you all know, the European Community is now coming together—in a way that no one believed possible even five years ago—because they, too, understand that the Cold War is over. Perhaps their amalgamation had something to do with convincing the Soviets and the others in Eastern Europe that they had to change.

To put it in its simplest terms, the world is changing dramatically and radically before our very eyes. Too often we are so caught up in what we are doing here that we do not understand the nature and the immensity of the change. I am convinced that the competition of the future will be more economic than it is military. If it is to be, and if we are to continue to be in the position where we are a leader in that competition, we have to change the way our economy is structured, and the way we relate to economic growth.

Your discussion here today is a very important part of that because you must help us move toward that change. I believe the bill that I have described is a small part of that. Much more must be done, and the work that you do, the interest that you have, the responsibility that you have, will be critical in bringing about change so that we can continue to be a leader in the kind of world that we are in today. Thank you very much.

MR. EIZENSTAT: We have time for one or two questions. Please take the opportunity and use the microphone.

PARTICIPANT: What does the acronym stand for in your bill?

CONGRESSMAN GEPHARDT: The Advanced Civilian Technology Agency.

MR. ZAININGER: Sir, what was it that made you all of a sudden understand the change in Europe so much better than before? I did not quite get that.

CONGRESSMAN GEPHARDT: Well, you know, we all read the newspaper and have an intellectual understanding of it. It helps, I think, to go and talk to Europeans who are in the middle of it and get their sense of importance, urgency, about what is in front of us and what can be done about it. You know, in our country we probably talk more about, think more about, care more about, for instance, the START talks because that has to do with missiles aimed at us and our missiles aimed at the Soviets, and we have, for a long time, been caught up in arms control that had to do with missiles rather than troops or tanks.

The Vienna negotiation is, of course, about Europe and it is a multilateral negotiation. It is not just the United States and Russia. It is everybody else. And so it helps, you do not read much about it here, frankly, you do not hear much about it. We hear about START, we do not talk about Vienna but when you go there and see it in process, you talk to the Europeans and they let you know how important and earth-shaking this is to them, and how the map is being redrawn. When they say it is the most important thing that has happened since World War II, you begin to get into their mindset and understand how vastly important it is.

In the last month since the Gorbachev changes that got the Vienna talks started, most of the talk in the United States is, yes, that is great but can you trust them? Will Gorbachev last? What I learned there in the last few days is, it does not matter if he lasts. Sure, we would like him to last, we hope he does, but if they destroy 40,000 tanks and lots of airplanes and lots of guns and lots of personnel carriers, even if Stalin returned, they would have a heck of a time building that armament back. We are literally asked in the treaty to do nothing, we get rid of virtually nothing. It is asymmetrical. The Russians and the Warsaw Pact countries are doing all the reduction. As the Russian negotiator said to me, "Do you people really understand what this means?" They are kind of perplexed that we are not moving a little faster.

MR. EIZENSTAT: Dick, I will make this the last point. You may be interested to know that in some of the earlier panels, particularly our corporate panel, there was a rather, I would say, cruel view of the capital gains cut. I think you might have been surprised by that. Do you have any sense of how that fits into this whole picture and where it is going to come out?

CONGRESSMAN GEPHARDT: Sure. Well, without giving a long talk on tax reform, let me just say that starting when I came to the Congress and through 13 years of service on the Ways and Means Committee, which ended Friday when I resigned because of my new position, I came to the conclusion that we had confounded our economic policy and outcomes by too great a reliance on the tax code to try to drive economic behavior. We had put so many gizmos in the code to try to get this to happen or that to happen or the other thing to happen, that we were causing people to lose interest in economics and to be interested only in tax breaks. That was the genesis of the Bradley-Gephardt Bill, and we were fortunate to have a President at the time who agreed with it, in principle, and we went ahead and I think wrote very good—although not perfect—tax reform legislation in 1986.

Just a side anecdote, in Europe last week, talking to business people, a lot of them were saying, yes, Europe 1992 is great and so on, but one of the things we have got to do in Europe is get these tax rates down. Even though you are being told that it is cheaper to get capital in Europe than it is in the United States, the Europeans say, that is really wrong, because we have got tax rates of 70, 80, 90 percent.

The only way we got our tax rate on high income down from 90, which is where it was when Kennedy was president, to 28 percent—or 33 if you understand that crazy bubble that I have decided never to talk about again—is by restoring the base. That means you take out some of the gizmos, the loopholes, the incentives, however you like to characterize them, which we littered the code with for over 40 years.

Now, when we got down to the final act, as is always the case in a democracy, you have a giant compromise. The compromise was that we would put up with this anomaly of letting the wealthiest have a lower rate than the people just below them, which is crazy. We would put up with getting rid of industrial development bonds and the investment tax credit and lots of other things that a lot of us really thought were pretty keen ideas, in order to get capital gains at the same rate as ordinary income, and to get the whole rate down to about 30 percent, which we thought was real progress.

Now people could invest, people could make decisions based on how they thought they could make money, not by some attractive gimmick in the tax code. The problem, and the great reaction that I and others have had to the President's suggestion, is that before the ink on the bill is dry, we are now back to trying to get a lower rate for capital gains. To make it worse, this particular proposal in the House was a two-year deal. It only gave it to gains for things that you had already decided to do, which makes no sense whatsoever. It is not even an incentive to have you invest. It is a windfall, entirely.

Again, it broke the symmetry, and, believe me, if it goes in the code and I suspect that it will, we will wind up a year from now with the investment tax credit people coming back. We will wind up two years from now with others coming back and you will be back to a 50 percent rate and a corporate rate to match. That, to me, is not progress and is not moving us in the right direction.

So that is all the fire over the capital gains, and you know, again, none of this alone is going to make that much difference. We have got about 12 or 15 things we have got to do at the same time, and we have got to do them well. Not any of these battles or skirmishes are going to win or lose the war, but we are in a war, an economic competition war. It is a heck of a lot better war to be in than the ones that we have been in in the past, but we have to win our share of this war.

We have to win some battles, and we have to be strong, not only for our kids and their material comfort, but because we are a leader in this world and other countries look to us. If we are borrowing 30 percent of our capital from somebody else, and we are unable to help the Polands when they come along, then that is a sorry situation and it does not need to be that way. We are better than that and we can get this done. Thank you very much.

MR. EIZENSTAT: Dick, thank you. You really added immeasurably to our discussion and Mike Wessel, we appreciate your help as well. Thank you again.

Gregg Jarrell will be our next speaker. He is a Professor of Economics and Finance and Director of Managerial Economics for the Research Center at the University of Rochester. He is also Senior Vice President at the ALCAR group, a Chicago-based software and management education firm where he served full time as Director of Research until 1988. He was the chief economist, as he mentioned during his question, at the SEC from 1984 to 1987. He has published several articles on the economics of tender offer regulation, and was a member of the SEC Advisory Committee on tender offer policy.

He graduated from the University of Delaware and the University of Chicago from which he received an MBA and Ph.D. Gregg will now give you a second bite of the apple, from his side this time.

DR. JARRELL: It is a pleasure to be here. I would like to thank Ed Abrahams for the invitation. I would like to thank Stu Eizenstat for the introduction, and I would like to thank my old friend, Commissioner Joe Grundfest, for essentially giving my speech last night at dinner.

I would like to start off by telling you a story which has a very strong moral to it. I work with a lot of lawyers. I get involved in a lot of these takeover battles, and the CEOs we've heard from, especially Mr. Booth, are a different crowd of people from corporate takeover lawyers and their advisers and investment bankers. A thing you have to remember about investment bankers is that they are just like people. You have to keep that model in mind when you try to understand how they operate.

I have a lawyer friend in Rochester and he told me this story which I hope you will find interesting. He said that he ran into a very unusual situation a few months ago. This elderly couple came into his office, and they were very elderly, mid-90s. They sat on separate sides of the room, and they had obviously been having quite a tiff. They announced to him tersely that they wanted his help in getting a divorce and he kept a straight face but he asked well, what is the problem, and they both agreed immediately that the problem was irreconcilable differences. They never really did like each other and they continued to make it impossible to get along with each other.

So he thought about this, and finally he said, well, you folks must be what, and the man said, I am 94 and she is 92. My friend said, well, you have obviously hung in there for a long time. Tell me, why now? Without missing a blink, they looked at him and they said, well, we thought we ought to wait for the children to die.

I like that story because of this discussion about short term and long term. Some people are willing to wait a very long period of time to get certain results that they view to be desirable, and other folks, in the modern age, if you don't get along that first week, you get divorced and you go searching around. I do not know who is right and who is wrong, but there do seem to be tremendous differences.

I come from the University of Rochester where the business school is called the William Simon Business School, named after Bill Simon who was our former Secretary of the Treasury. He is a big-time businessman, a wonderful guy, and that is for the record. Bill Simon got a very small fraction of the millions of dollars that he gave to the University of Rochester—to help us build a better business school, with an international reputation—from a deal involving the Gibson Greeting Card Company. It was one of the original firms that went LBO, and then within a few short years went public again at some phenomenal rate of return. It staggers the imagination. I do not even like to say it because I like Bill Simon and when you start talking about those kinds of riches, it is hard not to dislike someone.

He did quite well with that, and he has continued to do quite well after that. I think it is only appropriate. I am also good friends with Michael Jensen. In fact, I have his old job. Michael Jensen is now at Harvard. I am sure he would have been delighted to be here. He would have had a lot to add to the conversation. I talked to him the other day and he agreed that I should be his official spokesman. So anything that you wanted to ask Michael Jensen, you can ask me and he will have to live with the answer.

He started an outfit which I now am the director of, called the Managerial Economic Research Center. We have been doing some work that I wanted to share with you. It is very, very, very preliminary work but I think it has something to do with the topic today. One of the things we have done is to collect a sample of 160 big-time firms that have been restructured in the last four or five years, probably an exhaustive sample of all firms on the American and New York Stock Exchange that have announced in a fairly important way that they have restructured themselves or that they have embarked upon a restructuring program.

As many of you know, there are restructurings and then there are restructurings. There are some things that are restructuring in name only and you do it for one reason or another. There are other types of re-structurings that are major, that involve both sides of the balance sheet, levering up, paring down operations, making basic strategic decisions that are dramatically different from the path that you have been on. We tried a very simple way to distinguish between these two types of things. We developed a measure of the focus of the business.

Those of you in antitrust economics know there is a thing called a Herfindol index, which is one quantitative measure that in and of itself gives you a pretty good feel for what the industry is, how the industry is composed, whether it is made up of a lot of small competing firms or whether it is dominated by a small number of large firms. This Herfindol index concept we took to the firm itself in order to measure how far-flung were the firm's operations across the different businesses that they might operate in. A conglomerate would have a very low number, a firm that had all of their operations in one business would have a very high number.

We were able to develop this measure for each one of these firms on an annual basis. So by looking at this measure and this measure alone, we were trying to see whehter the firm restructured in a way that dramatically refocused its efforts, or whether it really just restructured in name only and did not change its basic strategic measure—at least, as reflected in this particular number. Then what we did was rank the 160 firms from those that had the biggest increase in this focus measure to those that had the smallest increase or the largest loss in this particular focus number.

We grouped them into portfolios. We said, okay, now, take the top 25 percent and put that in a portfolio. Take the bottom 25 percent and put that in a portfolio and measure the shareholder rate of return—capital gains plus dividends—over a long period of time, from the mid to late 1980s. The answer we got was that the firms that had refocused to sharpen their operations and the asset side of the balance sheet—completely ignoring the degree of leverage that they undertook, just focusing on business strategy— those firms had a rate of return that was, on average, 10 percentage points higher than the firms at the bottom of the list.

All of these restructured firms that refocused, on average, had a higher rate of return for their shareholders than others. I do not know what you make from this. I mean, if you do not believe in stock markets, you probably do not make much out of it, but it does indicate that there is something more than just debt that is going on here. The debt is not the entire story; leverage is not the entire story. You have to understand that some firms, when they restructure, must get in on that asset side and do something dramatically different.

We heard some talk earlier about Jensen's free cash flow theory and I wanted to say a few words about that because I think that it is a very simple theory and it is very controversial and it is very unproven. It is a very ingenious idea that strikes a lot of people as being intuitively obvious and helpful, yet it remains very much an untested proposition. If I could summarize it, I would start off by saying that Finance 101 is a simple subject. We tell MBA students and future business people in Finance 101 that your basic rule of operation is to look at projects and measure the rate of return that they are going to throw off. It sounds easy, right? You measure their internal rate of return. You estimate it.

If that internal rate of return for a project is greater than your cost of capital, then you do it. That is a go. If the internal rate of return is less than the cost of capital, then you do not do it. That is a no-go. Okay? That is what we teach them. It is a short course. The bright students say, well, how do you measure the internal rate of return of the project, and we say, that is a matter for statistics or quantitative analysis. What happens in many, many business firms is that the analysts try to figure out what projects the CEO is going to like, and for those projects they go ahead and say, well, that has got a good internal rate of return. Or there is some kind of mysterious process that goes on.

I do not mean to make fun of it. It is a very difficult thing to implement, and when you get all done measuring what you think the rate of return of the project will be, then you turn around and you measure the cost of capital. Well, good luck there, too. When I advise people on the cost of capital, I use a very simple rule. I say, look, rather than spend a lot of money and time and effort estimating your cost of capital, which you know you are going to get wrong anyway, just use the simple Jarrell Rule. The Jarrell Rule is this: if it is in a business that you have been in before and you have done well and you kind of think you know what you are doing, the cost of capital is 10 percent. If it is in a business that you have never been in before but you think you might be able to do okay, it is 20 percent. And if you are going into a business that you have never done before and you are not sure whether you are going to do well in, then it is 30 percent. It is that simple.

I have still not done my talk. What is free cash flow? Free cash flow is a result of having past successes in business. Businesses that have had past successes have investments and projects that are doing well. They are throwing off a lot of cash, and a lot of firms in this American economy are in that enviable position. They are throwing off cash. In the old days, they had to borrow or raise money. Now, they have prospered, they have done well, they are throwing off cash. The free cash flow problem arises when the stock market loses confidence in the ability of the senior management to "do the right thing" with the free cash flow.

What is the right thing? Well, go back to Finance 101. If you have new projects that allow you to get a higher rate of return on that free cash flow than the shareholders could get if you gave the money to them, then keep the money and invest it. If you do not have those sorts of projects, then give the money back to the shareholders somehow. There are dividends or stock buy-backs or—that is Finance 102, exactly how you do that.

So in order to be talking about the free cash flow theory, you must talk about a firm that has free cash flow. If the firm has got a free cash flow discount, you are really talking about a firm where the stock market does not have a lot of faith in management's ability to follow that simple Finance 101 rule.

Now, if the firm is subject to this free cash flow disease, it will have what is called a value gap. There will be a difference between the stock price out there in the open market, and the stock price which could prevail if somehow you could solve this free cash flow discount. This is Jensen's theory.

What is the role of debt in this theory? Here is the story. If you have a firm that is suffering this free cash flow discount, they are takeover bait. A raider can come in and buy it at the low price and then do whatever has to be done to solve it and make a windfall profit from the increase in value.

How does the incumbent management get out of this problem? Simple. This is a theory now. Incumbent management gets out of this problem by restructuring the liability side of the firm. What it does is it takes on a lot of debt, levers the firm up and then retires equity. What does that do? That eliminates the free cash flow. You have got no more free cash flow. All of your free cash flow, which was this excess cash coming from past investments, is now used to pay off the debt. It is like the old story where you are tying yourself to the mast of the ship so you don't do something that later you will be sorry for.

It is essentially what the theory says, that you limit the room for discretion on the part of management because Wall Street has lost confidence in management's ability to take that money and do the right or productive thing with it. Now, we know a lot of managements that may do this. A lot of the conglomerate mergers in the past have perhaps been a result of this. Professor Mueller is an expert in this area. He was one of the people that was screaming about this years and years ago. Perhaps this is one of the reasons for that type of behavior.

But one of the things I want to point out is that this free cash flow solution, this role of debt, is perfectly consistent with what people are concerned about in this conference. What people are concerned about is that debt constrains the ability of management to invest in something other than the interest payment on the debt. Here is somebody from the University of Chicago who is telling you that that is perfectly consistent with Jensen's theory. In fact, that is the inherent logic of the theory. They go hand in glove. It is supposed to constrain management.

Where does that leave this debate about research and development? It is consistent with the theory because if you have to make these interest payments, you are not going to be able to make all kinds of other discretionary expenses. Will research and development necessarily suffer? Sounds like it. It turns out it does not have to suffer, but if research and development were excessive, then it would be cut back. Presumably these managers would cut back on R&D. They would cut back on any kind of expenditure that they had made that was excessive. That is the logic of the theory.

This implies several things that are important. One, it implies that LBOs are not for all firms. They are for firms that suffer from this particular disease. Although at times, when you read *The Wall Street Journal*, you would think that LBOs and leveraged restructurings were taking over America. We have been dramatically surprised at the number of industries and the types of firms that have gone in for this type of high-leverage financing.

But this theory, in and of itself, implies that it is certainly not for all firms. It is only for those that are suffering from this free cash flow disease.

Now, let me continue on this theme for a minute. I am going to make a full circle. This sounds like a bad segue but I swear it is not. There is a tremendous amount of competition in the takeover market these days. The last two or three years, the competition has gotten very, very fierce. There are several reasons for this. One is that the defensive tactics that takeover targets can use essentially allow them a lot of resources and time to get an auction going. The regulatory officials, the courts in Delaware, everybody is very much in favor of the auction solution to takeovers.

There is some evidence on the horizon that this is changing. The Polaroid defense is the first break in this system whereby once you are in play, you are going to come out of it with high leverage. I do not want to get sidetracked, but the courts in Delaware appear to be trying to carve out some room for an in-between solution, and the Polaroid case is probably the best example of this new policy on the part of the Delaware courts.

But suffice it to say that there is a tremendous amount of competition and that means that bidders that win are paying top dollar, absolute top dollar. There are some staggering premiums that are paid. In my back yard, Kodak bought Sterling Drug, and if I recall correctly, on the day the market got the news that Kodak was "the winner," Kodak stock dropped 19 or 20 points. Kodak is bigger than Sterling, and that is a lot of money.

In fact, the stock market acted like Kodak bought this enormous, valuable drug firm and then on the way home dropped it and broke it.

The stock market wrote off most of the value, not the premium. They did not write the premium off. They went way beyond the premium. They wrote off a tremendous amount of value from Kodak and I do not think there is any evidence so far that Kodak has recovered from that. I do not know if that is good or bad. It is just that bidders are paying top dollar, and there is a thing out there called the winner's curse. Again, Professor Mueller knows about this.

There is also a tremendous amount of capital. Some have referred to it as predatory pools of capital—we heard that earlier today—enormous pools of capital looking for profitable outlets. Managements are bidders on almost all of these cases. They come up with a bid. Unions become bidders, employees become bidders. There are a lot of people bidding for these firms once they get into play, and there are a lot of people who are initiators willing to put them into play. So this is an enormous competition, and what is bothering people, what concerns policymakers, is that when firms get out of this fray, people are paying top dollar, they are financing a lot of it with debt, and they are getting closer and closer and closer to this razor's edge. Combine that with the fact that we have been in a prolonged business expansion. We have not been slapped around lately in economic terms. We have not been disciplined lately. We have been in this long economic expansion. Demand curves just keep moving to the right. Oh, they flit back a little bit but they just keep moving to the right.

I should also add that these high-leverage techniques are spreading to greater and greater industries that are more and more cyclical. It used to be that only the mature, low-cyclical industries were the LBOs. Don't worry, we thought, the LBOs are only being done there and they are temporary. Now, they do not look so temporary. Firms like KKR are hanging in there and the high leverage is hanging in there. Also, you have industries that are more cyclical having LBO activity in them.

What everybody is concerned about is that one of these days, demand is going to back up. We are going to go into this recession, and even with a fairly soft landing, they are worried about these high-levered firms littering the financial landscape and then having certain kinds of domino effects. That is what is bothering everybody. You do not want to sort of be the person that goes out and kills the golden goose. On the other hand, if you are a policymaker, you are paid to be concerned about those sorts of disasters. You are paid to be on the lookout for tomorrow's whoops.

In a junk bond market, what I am here to say is that I, too, share some of these concerns. Even Chicago economists have feelings, it is true. It is also a very cheap word, concern. It does not cost a nickel. Over the next two weeks, for those of you who are in Washington, just count how many times you hear the word concern. Count it. It is the most incredibly used word. It is so cheap. I was in Washington a week before I learned that. Whenever you are asked a question and you are tempted to say, oh, that is the most foolish question in the world, have you no faith in markets whatsoever? That is what you are saying inside your head. What comes out of your mouth is, that remains a top concern of the SEC and we continue to look at it with grave interest.

MR. EIZENSTAT: Now you took my talk.

DR. JARRELL: Anecdotally, there is a tremendous amount of support for this view that LBOs and hostile, high-debt takeovers and restructurings reduce R&D. You could put a line in here of distinguished people who know what they are talking about, been in business all their lives, who will swear to you on a stack of Bibles that that is true. I have anecdotes myself. I have been on the inside of these deals and I have seen on the bidder side, bidders that are budgeting cuts in R&D in order to pay for the financing over the next two years that they used to take the firm over. They will not say it under oath. They will say that is just a scenario—"We ran all kinds of scenarios"—but you know that that is what they are going to do.

On the defensive side, you see target management that would not have thought of cutting R&D over the last 10 years, but if they have got to do a leveraged restructuring to get out of this takeover pickle, what are they going to do? The first thing they do is, "We can cut R&D $10 million. That will give us a little bit of breathing room. That will finance some debt, right?"

One of the things that I was told at the University of Chicago is that the plural of anecdote are data. The data are that the Flamm overview and the Grundfest talk and the work that Bronwyn Hall and others have done indicate that it is very, very tough to find any support at the aggregate data level for the notion that R&D is suffering at the hands of the restructuring community. I was moved when I heard both gentlemen from the corporate community this morning refuse to condemn restructuring activity per se. Obviously they have some misgivings, and they have some concerns, but they refuse to condemn restructuring per se.

So I do think that one of the things we ought to come out of here with is there is not a real good basis upon which to go around attacking restructuring per se as a way to save the world in order to boost R&D. How come LBOs do not sacrifice research and development? It seems obvious that they could. Why don't they? I was surprised. I understand mergers in general. They tend to occur in low R&D industries and it is tough to find them, all right, but if you zero in on LBOs and look at 40 LBOs of big firms over the next three years, they have got to have decreases in R&D. But it does not show up; R&D as a fraction of sales does not fall.

But if you do not see it, why not? I think the answer must be that there is more to LBOs than just R&D and meeting debt payments. There is a lot more to them. There must be some sort of productivity increases in the cases that we are looking at that allow them to make the interest payments, to increase profits, to give that 30 percent to those avaricious, well-heeled investors that have put the money into this in the first place.

And another answer is that R&D is not sacrificed. A, it does not have to be in the LBOs that we seek because they are doing well, and B, the R&D can be justified. There is probably not an LBO on record where they did not change dramatically the composition of R&D. R.J. Reynolds, I think we have seen evidence of that in *The Wall Street Journal.* Didn't they cut out the smokeless cigarette? The first thing they did, right? Is that to say they cut R&D? They probably did in that firm, but who knows? Maybe it is the composition of R&D; I think that that is where the debate should go.

It is the composition: what are we investing in? The aggregate data are fairly uninteresting. We have got an R&D crisis. What are we investing in and why are we investing in it? It is a very complicated question.

I want to close with one more story. This is a story I am going to tell on my good friend, T. Boone Pickens, Jr., who is the famous would-be takeover specialist. He was a villain when he was taking over American oil companies or attempting to take over American oil companies and now he is a hero because he aimed his guns at a Japanese firm. He is a clever man.

I am telling this story just because I love to do his accent. When I was at the SEC, you know, you take a job like that not for the money. You take a job like that because you want some kind of recognition or press or to meet some people and have some fun in Washington. I did it for that reason as well, and one of the big, exciting things happened in the middle of the Unical–T. Boone Pickens–Mesa Petroleum takeover battle.

I had written an article, or actually Ken Lane, who is now the chief economist at SEC, wrote an article and I was helping take the credit for it. He wrote an article on this subject of long-term investment and myopia, is it true or is it not. I think that we were among the first people to look at this. Bronwyn, you were involved in it at the time. Pickens called me up in my office; this is when I first met him. I wrote an editorial for *The Wall Street Journal* and he read that and he got excited because he was in the middle of this battle. He did not have one friend in Washington at the time, so he called me up and congratulated me for the article and all this sort of stuff and he said he was going to be on TV that night on one of those talk shows. It was the Ted Koppel show and it was one of these debating things where they had T. Boone Pickens in one booth with the TV camera and they had Fred Hartley, the chairman of Unical, in another room. They were not actually with each other. They would have pounded each other to death. They were very much antagonists at the time. Separate rooms, and I thought, this is going to be wonderful. Pickens is going to mention my name and my research project.

So they had this debate and they are going back and forth and they cut for a commercial and Ted Koppel says, when we come back after a commercial, we are going to get to this issue about the myopic stock market, short-term, long-term. I thought, this is it. Of course, I had all my friends and relatives watching TV for my name, it was very exciting.

They came back from the commercial break and they went to Fred Hartley, and Fred Hartley said some things. Then they went to Mr. Pickens and they said, Mr. Pickens, you heard Mr. Hartley. He said that you are out for the short term, that you are only in this for a quick profit, that you are greedy and avaricious and that you are not in it for the long term. Did you hear all that?

And Mr. Pickens looked at Ted Koppel and he said, "Fred Hartley said all that, didn't he?" Ted Koppel says, "Yes, you heard it, didn't you?"

He says, "Yes, I heard all that. Was he on TV when he said that?" Koppel says, "Yes sir, we just had him, he was just saying it."

"And American viewers saw him, on the TV?" He was really confusing Ted. He says, "Yes, Mr. Pickens, they saw him."

"Well, then, they got a good, close-up view of Mr. Hartley, didn't they?"

"Yes, Mr. Pickens, they did get a good close-up view of Mr. Hartley. Could you please respond to this question?"

Mr. Pickens says, "Well, then you saw Mr. Hartley is not in very good shape. He is all puffy-faced and he has got a red nose and he is overweight and he does not take very good care of himself. Look at me. I stay in shape. I eat fresh fruits, cereal, I play racquetball every day. I am lean, I am mean. Look at Mr. Hartley there." He says, "Hell, if I was Fred Hartley, he said that about me? I am in it for the short term and he is in it for the long term? Hell, if I was Fred Hartley, I wouldn't buy green bananas."

That was the level of the debate at the time, and I am happy to say that the debate has elevated itself dramatically. Thank you very much.

MR. EIZENSTAT: Gregg, we appreciated both your wisdom and at this late hour, your humor as well. I must say that in listening to your description of the free cash flow theory and the limits that it puts on management, it reminds me on the public policy side of a policy some allege the Reagan Administration participated in, in which taxes were cut so deeply it created a deficit so large that it constrained members of Congress from spending anything that they wanted to spend. Whether that is actually a public policy analog I do not know, but it sounds at this hour about as close as I can come.

We will conclude, except for my very brief summary, with Dennis

Mueller who is a Professor of Economics at the nearby University of Maryland, in College Park, where he has taught since 1977. He received his Ph.D. from Princeton and a B.S. in math from Colorado College. His honors, like his publications, are too many to cite and still leave time for him to speak but I will mention just a very few. He has been the President of both the Public Choice Society and the Industrial Organization Society, which are separate fields in the economics profession. He received a Fulbright in 1987. He has written numerous books with titles that show his wide-ranging interests: *Public Choice*; *The Determinants and Effects of Mergers: An International Comparison*; and many others. I am literally omitting six pages of articles on his resume. We appreciate you coming. You have the opportunity to either look at yourself as the last speaker or the clean-up batter. I will call you the clean-up hitter. Thanks.

DR. MUELLER: Thank you. The one advantage I can think of going last is you do have the benefit of everyone else's comments and discussions so you can try and relate your talk to them rather briefly. The disadvantage, as Gregg so amply illustrates, is that most of what you wanted to say has already been said, and the greatest disadvantage, of course, is that it is so late in the day nobody really cares what you are going to say anyway.

What I am going to do first of all, so as not to prolong debate, is to agree with both Gregg Jarrell and Ken Flamm on what in some sense is the question of the day. That is, can we say anything about the relationship between leveraged buyouts and R&D expenditures in 1987, or maybe 1986 and 1987, or whatever the period is? I am going to agree and say I do not think we can say anything that is very definite, and I am not going to really talk about that very much.

Instead I want to talk about what I think ought to be our concern at a conference like this, namely more broadly based questions about corporate restructuring, again broadly defined to include mergers and acquisitions, hostile takeovers as well as LBOs and innovative activity.

Let me begin with the latter and ask some questions about what we do know about innovative activity. Where do innovations come from? What are the most innovative firms? First of all, we know that innovation is a very information-intensive activity. Information-intensive activities demand rather flat, decentralized organizational structures, with considerable expertise on the part of the top decision makers with regard to the technology of the firm. Innovative activity also entails greater risks than other forms of corporate decisions and thus requires more entrepreneurial leadership and an organizational environment in which risk taking is encouraged and rewarded.

All of this suggests that innovative firms will be relatively small and undiversified. (As an aside, I also suggest that proposals for large consortia, in general, are not going to be a solution to our problems.) Now, this image of relatively small, undiversified firms as being innovative, is an image that fits what we know about the sources of invention and innovation in this country. A disproportionate share of important inventions and innovations in this century has come from independent inventors and small firms— Polaroid is a classic example.

Even among the very large firms, the ones one can think of that have good, innovative records, the ones one can think of that have not lost their domestic markets to foreign competition and have even gotten good export performance, are typically companies like Boeing and IBM, some of the pharmaceutical firms and so forth that have stuck to their major lines of activity and have not engaged in a lot of acquisitions and diversification. Mr. Wendt of SmithKline Beecham I think would put SmithKline Beecham in that category also; the one large diversification effort they made when they acquired Beckman Instruments does not seem to have been a big success.

I am reminded in all of this of one of the characteristics of best-managed companies that Peters and Waterman uncovered in their study of 10 years ago or so. They termed it "they stick to their knitting"—they do not engage in much merger activity, and certainly very little that is unrelated to their major lines of business.

I do not think that this is just a coincidence. Rather, growth by internal means and external growth through acquisitions are substitutes. Let us look again where I began getting my spurs. If you go back to the first page of my vitae, you will see that some of my spurs were earned looking at conglomerate mergers back in the 1960s. Had I been precocious enough, I would have even named what I was doing a free cash flow theory of mergers and gotten some of the accolades that Mike Jensen is now getting for that.

We know that most of the active acquiring firms in the mid-1960s were based on mature, slow-growing industries: tobacco, textiles, paper, et cetera. These companies had basically three choices. They could continue to grow slowly and indeed decline relative to other firms in the economy, paying out their cash flow to stockholders to be reinvested elsewhere in the economy. They themselves could invest in R&D and invent and develop new products, or they could buy up other firms. They chose the latter route. The same choices faced the oil companies in the mid-1970s when the OPEC oil price increase swelled their profits, and many of them also opted for the acquisition alternative.

Now, how did these acquisitions turn out? The answer to this, it seems to me, depends on whom it is that you ask. I am not exactly sure who these people are any more, because the cast of characters seems to be changing. The last time I was at a conference like this, a mixture of academicians and business, and Wall Street type people, where Michael Jensen was present,

I went there assuming we would be on opposite sides of the fence, and disagreeing about all this, and Michael Jensen got up and seemed to be saying things that had a sort of deja vu ring to them.

At that time, I thought Gregg Jarrell and I were disagreeing on these issues. Today, I thought I heard him saying that acquisitions by Kodak and so on were not wealth increasing. But it used to be that there were people from finance departments, and indeed a very famous (and very often cited) survey by Jensen and Ruback, that basically said that the mergers of the 1960s, conglomerate and otherwise, were a success, as were the mergers of the 1970s and 1980s. If you go back to the Jensen and Ruback paper which was, I think, part of this empirical consensus that Kenneth Flamm talks about as to mergers and takeovers having been on average a success, you will see that most of the studies use databases heavily weighted toward the 1960s.

Why were they a success? Because on average the premium paid for the acquired firms was greater than the losses, if there were losses, to the acquiring firms at the times the mergers were announced.

Now, some of you may be surprised to learn that the empirical consensus is that the mergers of the 1960s were a success. Indeed, a consensus of writers for such publications as *Business Week, Fortune, Forbes, Wall Street Journal*, et cetera, would probably run the other way. Indeed, since many of the so-called bust-up takeovers of the 1970s and 1980s were motivated to undo earlier mergers, one might even assume that at least one set of mergers must have been characterized as mistakes.

But such is the beauty of the market for corporate control, at least in the eyes of some of its ardent champions. I guess at least the two people from Wall Street this morning would still be in that camp, maintaining that wealth is created when companies are put together and wealth is created when companies are broken apart. Is it any wonder, therefore, that the criticisms of takeover activity heard from its champions, let us say, on Wall Street, are only that we have too little takeover activity, that the Williams Act slows up takeovers and thus stops this wealth-creating process and that, for example, section seven in the Clayton Act, when we used to enforce it, prevented too many synergy-enhancing, horizontal, and vertical acquisitions?

Now, this chorus of enthusiasm is not unanimous and the reason *Business Week* and *Harvard Business Review*, et cetera, often seem to come to alternative conclusions is that they do not judge mergers' success entirely on the basis of stock market reactions at the time the mergers are announced. Instead they look at how the merging companies perform in the months and years following mergers. Similarly, economists who have studied the real performance of corporations, namely, their profits and market share over time—and I am thinking here of Professors Scherer and

Caves at Harvard as well as myself—we have found that when you look at the performance of merging firms and particularly acquired companies in the years following acquisition, that the performance tends to be, on average, negative in terms of profit declines and market share declines and the like.

One of the studies I did was of acquisitions in the 1950s and 1960s. I tracked the acquired firms for an average of 11 years after they were acquired. I found a significant destruction of wealth following the firm's acquisition. A typical acquired firm was, in fact, projected to basically disappear after 20 years.

It is worth noting that these were not simply firms that were on their way out anyway. Included in this group of firms were companies like Harley Davidson in motorcycles, Talon in zippers, companies that, in fact, were the industry leaders prior to acquisition and then subsequently were displaced by Japanese firms. Other firms included small and very up-and-coming firms like Hamms Brewery and Vapor Valve and Pump and a bunch of others whose growth trajectories were basically turned around following their acquisition.

On face value, it is hard to envisage why hostile takeovers would improve the innovative records of the companies involved. Certainly, the immediate effects of hostile takeovers and the restructurings that follow— and of leveraged buyouts, for that matter—and the pressure to increase immediate earnings, are unlikely to create a corporate environment that enhances the long-run commitment to R&D and investment and risk taking that the innovative activity requires.

In thinking about these issues, it is perhaps useful to look at a concrete example, namely, the machine tool industry in the United States. In 1975, the U.S. was the world's largest producer of machine tools, was the second largest exporter, and it had the lowest dependency on imports. In 1987, 12 years later, we were fourth in production behind Japan, West Germany and the USSR, sixth in exports behind countries like East Germany, and more than 50 percent of our purchases were in the form of imports. What is perhaps worst about this example is that we lost market share because we lagged in introducing innovations in a market in which we were the technological leader, at least back in the 1970s.

Here I might comment on Mr. Tokarz's assertion—it is easy to attack people now that they have gone on the Pan Am shuttle back to New York— that Japan had targeted machine tools as a market to attack back in the 1970s and that helps explain why we lost market share. I am very skeptical of that kind of "Japan bashing." My suspicion is that if Polaroid were a Japanese firm and had come out with the instant development camera in the last five years, there would be at least one congressman and two senators from New York coming down to Washington and saying that Japan had targeted our amateur camera market for attack and that in particular Kodak was being singled out for a ruthless attack.

If it was targeted, it was probably because it was a market ripe to be taken because of the lag of innovative activity by the firms in that industry. It is interesting when you look at that industry that among the leading machine tool firms were several growth-through-merger conglomerates: Litton Industries, Houdaille, White Consolidated, Textron, Bendix, Colt. And yet they fared no better and probably worse than the undiversified companies in that industry.

I had planned to talk about two firms as concrete examples in this talk, and as it would be, these have been heavily discussed already today. But let me perhaps give a little different light on these two companies. I think it is particularly interesting in this regard to look at the history of Houdaille, a firm which for basically a half century produced auto parts in this country. In the 1960s they went the "typical conglomerate" route; they started to lose their auto parts market and so they began conglomerate diversification. They moved heavily into machine tools and by the mid-1970s they were one of the leading machine tool manufacturers in the country.

In 1979, as everybody now knows who was here this morning, they went through a leveraged buyout and in the next six years, their constant dollars sales dropped by over 70 percent. Shortly after those six years were up, the entire division was liquidated.

Now, it is true that undiversified firms also suffered badly in this period. My point is not that mergers in some sense caused the particular problems of the firms in the machine tool industry. But certainly the mergers and the other kinds of restructuring that went on do not seem to have led to those firms outperforming the other companies in that industry. Indeed, it is inconceivable that Houdaille could have had a worse trajectory in machine tools than it had following the leveraged buyout.

Beatrice is another good example which was also discussed this morning. Maybe I can give a slightly different perspective on that. Beatrice was, in fact, the only exception in the Peters and Waterman book to the "stick to the knitting" rule. It was one of their best-managed companies, and yet it had engaged in a lot of acquisition activity, basically acquiring small firms that fit into its corporate mix of food-manufacturing companies.

No sooner was the ink dry on the millionth copy of *In Search of Excellence* than Beatrice itself became the target of a takeover movement by S-Mark (which was formerly Swift but which had also recently acquired Norton-Simon). Beatrice was forced, to avoid being taken over, to turn around and take S-Mark over itself. By the mid-1980s, *Business Week* was calling Beatrice an acquisition junkie and Beatrice was, in fact, spinning off divisions to regain earnings and to improve its performance.

Rumors of hostile takeover continued to circulate and eventually Beatrice went through a management buyout. I think if you are going to talk about restructuring, you have got to look at the entire 10-year period, including the mergers that Beatrice undertook, including the management buyout and so on. Certainly all of that period was not one in which I would guess Peters and Waterman, if they were to repeat their study, would have classified Beatrice in the best-managed group. Maybe the management buyout did improve Beatrice's performance relative to the period just before the management buyout, but there was a lot of restructuring going on both before and afterwards.

Well, if one listens to people who champion a free market for corporate control and hears only good news about the effects of these kinds of transactions, then conglomerate mergers of the 1960s increased efficiency and corporate wealth, and the mergers and takeovers of the 1970s and 1980s are doing so also. Eleven years ago I estimated that merger waves following World War II had resulted in some 25,000 companies being acquired through merger or takeover. In the intervening 11 years, we have probably had another 25,000 companies disappear.

Yet, despite all of the bad managements that have been ousted and the synergy that has been created by this merger and acquisition activity, the country continues to decline in productivity. We continue to lose our markets to the Japanese and other foreign firms.

As I mentioned this morning, for every 10 mergers and acquisitions in this country, there is one that takes place in Japan. One has to, I think, ask the question how it is the Japanese seem to remain so innovative and efficient without the tight constraints on the market for corporate control, and the discipline on managements which we get through the much heavier and more intense merger and acquisition activity that we have in this country. (Here I might mention, on the point raised this morning, that the Japanese can engage in cooperative R&D but, of course, the diversification mergers that have taken place in this country are supposed to have a similar effect. The synergies that were expected from them were because one division would be able to cooperate in terms of R&D with the other divisions and so on. So partly we were supposed to accomplish some of these things through diversification and the mergers that went with it.)

Now, obviously, correlation does not imply causation, and one cannot conclude from this that mergers and acquisitions have caused our poor industrial performance over the last 20 years. But I do think that these facts fit rather uncomfortably together, and I think it is about time that we begin asking some very serious questions—well, people have been asking, maybe it is about time we begin answering some very serious questions. How can it be that our macroeconomic performance, in terms of productivity and growth and so on, continues to go down—or certainly not go up—and yet on the micro level, we continue to conclude that mergers, acquisitions, hostile takeovers, and so on, are improving the efficiency of the companies involved?

Here, just a final note on demonstration effects. If there are demonstration effects to these takeovers and so on, then they should appear as an upward trend in efficiency and performance over time. Certainly, now, you would expect that in 1989 the demonstration effects of the market for corporate control and action would be that managers would be much more sensitive in 1989 to the threat of takeover; they should be doing a better job than they were doing in 1969. So we should begin to see an improvement in performance, in efficiency, through demonstration effects at the macro level, and yet I think it is hard to discern and I think we should begin looking for it.

I have some speculative suggestions for policy but I think I will close at this point because some of you, and certainly Stu Eizenstat, want to have a chance to talk.

MR. EIZENSTAT: Thank you. Are there any comments or questions for our panelists?

DR. LICHTENBERG: I think Dennis and perhaps some of the other speakers may have described slightly inaccurately the macroeconomic record, particularly on productivity growth. My reading of the trends on productivity is we had a dramatic slowdown in U.S. industrial productivity starting in 1973. However, during the 1980s, we have had some resurgence in productivity growth. Certainly, strongly in the manufacturing sector where we can measure productivity best. In fact, the productivity growth rate, starting in about 1980, is higher than it was even in the golden age of 1945 to 1973.

Now, in fact, that fits fairly well with the story that yes, perhaps the conglomerate merger takeovers were a mistake, they may have contributed to the productivity slowdown of the 1970s. In the 1980s we have accelerated the undoing of some of that damage so the time-series evidence does provide some support for the notion that the takeovers of the 1970s and the 1980s and the changes in ownership have, in fact, been consistent with productivity improvement.

MR. EIZENSTAT: May I just myself to that? My understanding is that in the manufacturing area there has been an upsurge in productivity. However, if you look at the overall productivity figures for the decade, they still are below 2 percent. They are not really terribly exciting, and the productivity has increased in precisely that area most subject to foreign competition, so you also could make the argument that it may not be the leveraged buyouts and restructuring so much as the downsizing of those industries and the requirement to mechanize and to have to compete with foreign competition and not have the luxury of having as many men and women doing the work as you did before the foreign competition was there. That is just a non-academic response.

DR. LICHTENBERG: I agree with you, I do not want to rely too heavily on the macroeconomic data. We need microeconomic studies, and there are some of those around which I think support the view of productivity enhancement.

MR. EIZENSTAT: Any comments on that point?

DR. MUELLER: This is not my field, I must say, but you can come back with about 14 different references on each side. I guess my reading of Martin Bailey's work, of Brookings' work, would suggest that productivity, even in manufacturing, has not bounced back. But apart from this, there is a major problem determining where you are starting these trends and where you are stopping these trends as to whether we have a serious problem in that regard.

MR. FLAX: We seem to have a lot of conflicting argument here, even coming from people on the same side. For example, we are told that what counts is that we are really not attacking industries that have substantial R&D. We are attacking Beatrices, we are attacking retail trade. Some of those are not successful either. There have been numerous others. These things have changed hands. What is happening, and I think it lies outside the area of this conference, but certainly has been happening, is the redistribution of income.

The reason people say that first conglomeration and then deconglomeration were equally successful is that they put money into people's pockets, and if you look at the statistics, they show that there is ever-greater concentration of wealth, something like what happened before the 1929 market crash. In this game of musical chairs in finance, there is always a new gimmick. I read that Mr. Milken says now that the junk bonds are not really appropriate for the present era. He is ready to go in some other direction if we give him the chance.

But let me say one thing more which is in the nature of a question. I have been told by numerous people, including Mr. Grundfest, that all I had to do was read the Jensen paper and I would find all the evidence I need. I do not find any evidence. I find a purely theoretical economic argument of the Chicago school. I recognize it because I used to work with Bill Niskanen, and it is right there. But evidence, no. The business of converting anecdotes into data and data into knowledge, which is another step, it is doctrinaire. So those who told me I would get the answer out of this particular paper have misled me, or I am too stupid to extract the answer. I wonder if any of you have seen the answer here.

DR. JARRELL: Who would have thought you would have read it? It is sort of an easy answer to give. You give somebody an academic paper and you say, well, you just read that and you will find the answer and usually that takes care of them. I would want to tell you that Joe Grundfest's office number is 272-2400, and that you ought to give him a piece of your mind.

But less frivolously, it takes a lot more money to write empirical papers than it does to write theoretical papers. Jensen is a clever fellow and he can rip off one of those things and it is a good idea and you get all this press; as Dennis well knows, you can struggle and work for literally years to produce a piece of empirical work and by the time you get finished, nobody cares. You know, three people read it, including your parents. So that is a very difficult area for our profession.

MS. HALL: I am Bronwyn Hall from the University of California at Berkeley and Gregg just gave me an excellent introduction. I am an empirical worker with some late-breaking facts. Ken Flamm, when he wrote his overview, did not have the latest version of my work in progress and I thought it would be useful if I gave you a couple of results.

The first one is small but it basically confirms his hypothesis. I had mentioned it this summer in the testimony I gave to Congress. It is that the difference between the NSF and my results at least—I cannot speak for Frank Lichtenberg—is that I am talking about R&D intensity when I ask does it increase or decline after acquisition, and they are talking about the level of R&D. There is no question about the fact that acquisition in general produces some downsizing in which sales decline and therefore the R&D intensity tends to stay pretty comparable, so that is the first answer.

MR. EIZENSTAT: Another theory that the sales dropped.

MS. HALL: So the relationship stays the same, which I take to mean that nothing much has changed really drastically at that level. On the other hand, it is true that there is downsizing going on, but that may be a good thing. I mean, you cannot say whether it is bad or good yet. You do not have enough information.

The second fact is a little more interesting. I looked at the thing he was asking about, which is, suppose you now ignore leveraged buyouts and look at acquisitions in the manufacturing sector which are public firms and which are leveraged at the time they take place. It turns out, and I do not know what to make of this, that if you ask, what is the R&D intensity of the firms that make these acquisitions, zero, one, two years later, it actually goes up relative to other firms in their industry by a tiny bit. Again, not significant, but it is certainly not the case that those showed declines either.

On the other hand, it is also the case that the acquisitions themselves are relatively small fractions of the firms which are doing the acquiring and I do not think you would expect leverage to have an overwhelming effect on that. So if you think there is just a problem caused by leverage, it is not there. It is not there in that sample. I think the next place to look is firms that leverage in order to avoid hostile takeovers or other types of attack, and that I cannot give you the answer on yet.

I also want to comment on the topic that Ken raised, which I raised two years ago or so, that he calls the demonstration effect. If everybody looks around him and sees that maybe he is a target, then the control sample, the firms that did not get acquired, their R&D behavior could be almost the same as those that were acquired, so that our tests are not very good. We cannot tell if the whole economy has reduced its R&D in response to what it sees as a more active churning of the capital market; we are not measuring that. I think that is an important question actually, and I just want to underscore that I agree with Ken, that we may not be doing the right tests yet, we empirical workers.

My third comment is for Gregg Jarrell. He alluded to a result on leveraged buyouts which is not really a result. It does not actually exist, although I think the spirit of what he said is correct. In leveraged buyouts, nobody has actually gone out and asked what happens to research and development in the vast total of leveraged buyouts after they take place. The only study that actually did that is the KKR study and they have only their own sample, which is small. I have something like 80 and, of course, I do not see the numbers after the leveraged buyout takes place. Steve Caplan—he is at The University of Chicago—has made some attempt to look at numbers after leveraged buyouts take place in about 50 cases.

The truth is it is hard to do. The reason it is hard to do is because there is not a whole lot of R&D in those firms to begin with and that is the important fact. It is not whether the R&D is going up or going down or whatever, it is that in leveraged buyout activity, even in the last three years, even in this most active period, the amount of industrial R&D involved is one-quarter of 1 percent of the total industrial private R&D spending in this country. You could just take it away and it would not have an effect. I mean, it is too small a number, and that is actually the important fact about leveraged buyouts and why many of us, myself included, are looking elsewhere now.

MR. EIZENSTAT: Thank you for that update. We will conclude with Dennis and then I will give my very brief summary.

DR. MUELLER: I just wanted to chime in on one of the points that was made. One was to interpret my results—that market shares of acquired firms decline after they are taken over or merged—as indicating that downsizing is occurring, and indeed if you wait 20 years, the average firm is downsized out of existence. That, in itself, is neither good nor bad. It depends on who takes up the slack, and I think Chicago-type economists like me tend to assume the market will work, and somebody else will come in and produce a better widget. I think what we have observed is the market does work and it is the Japanese that come in and produce a better widget.

That may or may not be something we want to be concerned about, but I think that is the kind of question we want to be looking at, and whether there would be a better way of changing incentives and controls so we would keep some of these firms—not have to downsize them out of existence—and maybe get them recycled and on the upward track.

Now, again, the market works and we'll find something to do, but there are parts of Europe where it almost looks like their main industry in the future is going to be tourism. I mean, they are good cooks and it is very pleasant to go there and the Japanese will go there for their tourist activities, but they are basically disappearing from productive activity. That is the kind of question—who is taking up the slack as these firms disappear—that certainly I do not have an answer to, and I doubt if any of us in this panel does, but I think it is a key question here.

MR. EIZENSTAT: Let me conclude with first some thank you's for all of the panelists for the time and effort and really very rich contributions they made. I want to particularly thank our two CEOs, Mack Booth and Henry Wendt, for their contributions and for staying. It shows a depth of interest in the topic which we really appreciate. I want to thank the Academy Industry Program, and the two National Academies of Sciences and Engineering for hosting this symposium, and I am told that there will be a very prompt publishing of a transcript, sent to all of you and widely distributed and rightly so.

This is obviously a difficult topic to summarize but let me make a stab at it. First, it does seem to me that there is a dramatically different Wall Street view from the corporate view. Quite obviously they have different economic interests. Wall Street makes money on it and it is an aggravation to corporate America. The academic data are obviously quite mixed at best. If anything, they certainly do not seem to tilt toward any evidence that there is a significant reduction in R&D from restructuring.

The Wall Street view is that the major problem we have in competitiveness in R&D spending—which everyone admits is a problem—relates almost wholly to the cost of capital, that restructuring, if anything, is positive, that privatizing these companies and giving the managers an equity share actually encourages better management. Indeed, it may encourage more R&D, although it is hard to know since acquired companies are sold within three to five years. In any event, that is their view. By their own admission, on the other hand, LBO candidates tend to be rather low-intensity R&D performers, and this may make a real analysis of the impact on R&D very difficult to come to.

The corporate view is that the real villain is the short-term horizon, and that the short-term horizon forces corporations to make short-term decisions, to satisfy stockholders who want to know what the next quarter returns are going to be, and therefore take away from the necessary concentration on R&D, which is obviously a long-term interest. In this view, LBOs are, if anything, essentially a negative to R&D because, again, they tend to force a short-term attitude of paying off the debt and then selling the company.

Again, the data do not seem to support—although anecdotally they might—the intuition one would have that higher debt levels for these LBOs would lead to less R&D. That again may be because we are dealing with firms with low R&D intensity. It may be, in fact, that these LBOs create so much more productivity and growth that companies can, in effect, pay off that debt, pay the rate of return that is required and still keep their R&D up.

For me, the most important thing, of course, is the public policy implications of this. It seems to me there are, perhaps, two, and I will conclude with these. Congress, number one, did take a look at takeovers, last year and the year before. There was a great frenzy in the Congress about takeovers. There were hearings in the Ways and Means and Finance Committees; there were hearings in the Energy and Commerce Committee. Basically, although Congress wanted to act because it sensed the public wanted action, it did not. It did not even pass a bill out of committee, and I think the reason it did not is it realized that the data did not support dramatic action to limit takeovers. There were, as you know, efforts to try to eliminate the deduction for debt in takeovers, which was an obvious way to handle the situation, but then somebody said, well, that is great, except when a foreign corporation comes in to do a takeover, and they do it through debt. You cannot stop them from deducting the debt off their country's taxes.

So every time Congress looked at getting to first base, it found three other reasons why getting to first base might actually be going backward rather than forward. One of the reasons that we do not see more congressional action right now is, I think, simply that there has not yet congealed an academic view. I am a great believer in the fact that we have an almost unbelievable open public policy and political system. The guys up there, the Dick Gephardts of the world, look to you to tell them whether or not what they want to do instinctively is the right thing to do. They knew what they wanted to do instinctively, which was to stop these takeovers, but they could not find the academic and business consensus on doing it and certainly on how to do it.

I think what this conference today has shown conclusively is that there are no conclusive data. That is an important point. It is sort of like a President not making a decision to fire a missile. That is as much a decision as if he fired it. It may be the right one.

Here, not having academic consensus, not having adequate data, it seems to me that we ought to be moving extremely slowly in anything that would limit restructurings, until we have better data. Maybe what we ought to do is get the Office of Science and Technology Policy, and the Congress with this one trillion dollar budget, to give somebody a million bucks to really do a first rate empirical study so we can know. We have volunteers. But the point is that we do not have that data now, and when you do not have them, you do not act because your actions may be worse than the status quo.

The second point and the last point that I would make is that it is not to say there has not been a consensus stated here. There is clearly a consensus that we have competitiveness problems; there is clearly a consensus that R&D spending is declining. Whether it is due to leveraged buyouts or whatever, that it is not—or at least its rate of increase is not—what we would like to see. There is clearly a consensus that capital costs are higher than our competitors whether 25 percent, 75 percent, 100 percent or 300 percent, they are still higher, and you have got to figure that when capital costs are higher than your competitors, you are at somewhat of a competitive disadvantage. And there is, in fact, built into our economic system and maybe into our tax system, a sort of tilt toward short-term horizons that forces our CEOs and our best managers to act in ways they wish they did not have to, and would not if they could be given some relief from these pressures.

I think we got some extremely provocative and very interesting notions from our corporate panel and from others about possible tax policies to change the short-term horizon problem: taxing pension funds, a graduated capital gains tax, ways to reduce our capital costs.

Perhaps until and unless we have the kind of empirical studies that will develop a consensus on restructuring, what we ought to be doing is focusing on those things where we have reached some form of consensus, which are: we need to spend a hell of a lot more on R&D than we are; there are serious reasons for the R&D decline; we do have short-term horizon problems; and we do have capital cost problems.

It seems to me that is a very productive eight-hour day that we all put in. I thank you for coming, I thank the Academies for producing it, I thank the audience for tolerating a long day. Thank you.

# Appendix A
# Industrial Research and Corporate Restructuring: An Overview of Some Issues

Kenneth Flamm[1]

On October 12, a conference convened by the Academy Industry Program of the National Research Council met to consider the significance for industrial research of recent trends in corporate restructuring activity. This brief overview essay was intended to raise some questions for that discussion.[2]

Are there grounds for concern with respect to the recent behavior of American industrial research and development activity? What is the available evidence on the relationship between recent corporate restructuring activity and industry R&D? What are the positive and negative perspectives on how corporate restructuring affects industrial R&D? Finally, can any tentative assessments of its impact be ventured?

## IS THERE REASON FOR CONCERN?

Since about 1987, it has been clear to both business analysts and public officials that industrial research in the United States has slowed considerably after almost a decade of rapid growth. Figure 1 shows both absolute levels

[1] Consultant, Academy Industry Program of the National Research Council. The author is a Senior Fellow at The Brookings Institution, 1775 Massachusetts Ave., N.W., Washington, DC 20036. The opinions expressed in this essay are the author's alone, and should not be ascribed to officers or staff members of the National Research Council or The Brookings Institution.

[2] Without implicating her in my errors, I thank Margaret Blair for her very helpful comments.

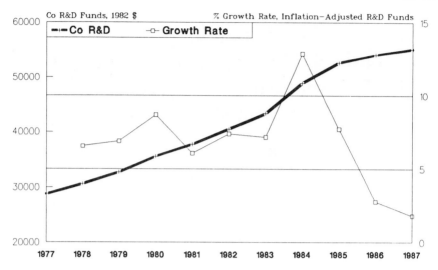

FIGURE 1    Company-funded R&D expenditures.

and growth rates for company-funded research and development funds as measured by the National Science Foundation (NSF).[3] For the period from 1977 to 1985, growth rates in inflation-adjusted company funds for R&D did not fall below 6 percent. Since 1986, however, the growth rate has fallen below half of that minimum, and forecasts suggest further decline in 1989.[4]

Why that R&D spending growth rate has fallen is less clear. One way of gaining insight is to examine the R&D-to-sales ratio, a measure of the *research intensity* of a company (or industry). Figure 2 shows the recent historical behavior of research intensity, as measured by company-funded

---

[3] The NSF figures on company-funded R&D used here include all company funds expended on the NSF's definition of R&D (which differs from that used in FASB-based accounting reports, *Business Week*, corporate annual reports, and many internal company cost accounting systems) on projects undertaken *within* the company's U.S. facilities. Excluded are company-funded R&D conducted by external organizations, company-funded R&D undertaken within foreign facilities, and federally funded R&D performed within domestic facilities.

The NSF data used in this overview are the most recent data available, unpublished statistics furnished to the author by the NSF. The precise firms used to estimate these figures, and their industrial classification, are based on the construction of a new panel for the NSF survey in 1987. Earlier data released by the NSF are based on a sample put together for 1981. For that reason, these data may not exactly match earlier industrial R&D data produced by the NSF.

[4] See National Science Foundation, *Science Resource Highlights*, "Modest Increase in Company R&D Funding Estimated for 1989," NSF 89-310, June 30, 1989; Gene Koretz, "Business Talks a Better R&D Game Than It Plays," *Business Week*, August 21, 1989, p. 20.

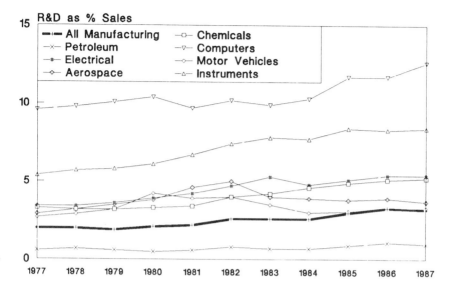

FIGURE 2   Research intensity of R&D-performing manufacturing companies.

R&D per dollar of sales, in key industries. Research intensity generally grew during the late 1970s and early 1980s, but with a few exceptions—the computer and chemical industries, in particular—stabilized or even dropped slightly in the mid-1980s.

The percentage rate of change in inflation-adjusted R&D spending can be broken down into two components—change in research intensity, and change in inflation-adjusted sales.[5] Figure 3 shows that the declines in 1986 and 1987 in inflation-adjusted R&D did not reflect a continuous decline in either research intensity or inflation-adjusted sales of American research-performing firms in manufacturing.[6] In the face of a declining sales base in 1985 and 1986, the relative size of the research effort of American R&D-performing companies increased. When sales picked up in 1987, the intensity of these companies' research efforts decreased.

Breaking out growth in inflation-adjusted company R&D funds by major R&D-performing industry, a similarly heterogeneous picture emerges

---

[5] If $R$ is R&D, $S$ is sales, $P$ is the price index used for inflation adjustment, and $ln$ ( ) denotes taking the natural logarithm of some argument, we have:

$$ln\ (R/P) = ln\ (R/S) + ln\ (S/P)$$

Take first differences on both sides of this equation, note that first differences of natural logs are approximately the percentage rate of change, and we have the procedure described here.
The GNP implicit price deflator is used for inflation adjustment.

[6] Manufacturing firms were estimated to account for about 91 percent of U.S. company-funded R&D in 1987.

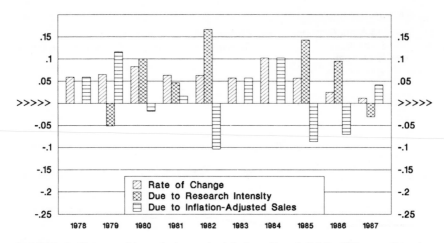

FIGURE 3 Decomposition of change in inflation-adjusted R&D; U.S. manufacturing, company funds.

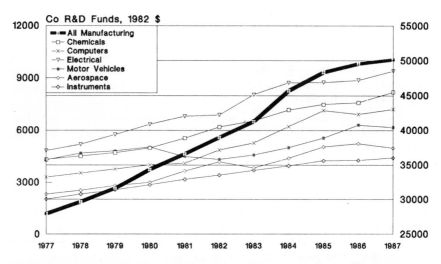

FIGURE 4 Growth in company R&D funds, by major sector. Left axis: all manufacturing; right axis: other sectors.

(see Figure 4). While there was little growth in such major sectors as computers, motor vehicles, aerospace, and instruments, in 1987 chemicals and electrical equipment grew respectably.

Is there some deeper logic to the overall sluggish growth that emerges from the aggregation of these seemingly diverse industry experiences? The importance of the issue seems obvious enough: R&D investment is central to the long-term competitiveness of high-technology industries, and those high-technology industries are at the heart of any strategy to increase American living standards. The fact that the National Science Foundation had received numerous telephone inquiries concerning the impact of a wave of corporate restructurings, which became highly visible around 1983, apparently led it to examine the impact of restructuring on R&D.

## THE RISE OF THE LBO

Corporate restructuring is a vague term which encompasses many types of activities. It is often used to describe merger and acquisition activity. In the United States, a surge in corporate takeovers in the 1980s was especially notable; this activity coincided with the rapid diffusion of a major financial innovation, the so-called leveraged buyout (LBO). Within the science and technology community, however, other events, like the divestiture of the regional operating companies by AT&T in 1984, are also included in discussions of the potential impacts of corporate reorganization on research investment in the United States. The creation of the AT&T Bell Laboratories and Bell Communications Research from the predivestiture Bell Labs; the transfer of RCA's Sarnoff Laboratories to SRI International in the wake of General Electric's acquisition of RCA (and sell-off of its consumer electronics business); and the reorganization of research at G.E.'s Schenectady central research facility are examples of possibly significant "corporate restructuring" which are not directly linked to the high-profile leveraged buyout.

The rise of the leveraged buyout was a relatively recent development on the American financial scene. The use of so-called "junk bonds" (low-quality bonds) to finance takeover activity was considered a significant innovation in financial markets when it first became widespread in the 1980s. LBO activity appears to have been virtually nonexistent prior to the early 1980s. Data cited by Lichtenberg and Siegel show LBOs rising from under 4 percent of the value of all merger and acquisition deals in 1981, to about 27 percent in 1986.[7]

What defines a leveraged buyout? It is helpful to think of "leveraged corporate takeovers" as a subset of all acquisition activity, in which a group of investors utilizing significant borrowed funds buys out a controlling interest in a publicly held corporation. "Leveraged buyouts" are generally understood to be the subset of leveraged corporate takeovers in which a group of investors buys out *all* the outstanding equity, and takes a company private, while making use of substantial borrowed funds. What constitutes

---

[7] Frank R. Lichtenberg and Donald Siegel, "The Effects of Leveraged Buyouts on Productivity and Related Aspects of Firm Behavior," NBER Working Paper 3022, June 1989, Table 1.

"substantial" or "significant" use of borrowed funds evades precise definition; the borrowing is generally considered to result in a "significant" increase in a company's debt-to-equity ratio. Hostile takeovers are conducted over the objections of current management, while friendly takeovers involve the acquiescence of managers. LBOs, like other takeovers, can be hostile or friendly. "Management buyouts" are LBOs in which current management participates in the acquiring investor group; such deals presumably are friendly by definition.

## EXPLAINING THE SLOWDOWN IN R&D GROWTH

More than telephone calls from the curious pointed the NSF's attention to the LBO phenomenon. The NSF's annual surveys of R&D directors at major corporations, and comments received in response to its spending survey, had suggested that research directors, as a group, often perceived the impact of corporate restructuring to be a significant factor in cutting back research at their own and other companies.

In 1988, the NSF initiated an attempt to identify the impact of restructuring on companies responding to its annual expenditure survey.[8] The 200 largest R&D-performing companies (accounting for about 90 percent of company funds) were divided into two groups. One group consisted of firms uninvolved in any sort of merger activity. Another group consisted of 33 large firms which had merged into 16 companies, and 8 companies involved in "LBO and other restructurings" (which included leveraged buyouts, as well as stock buybacks and other restructurings to defend against leveraged buyouts) over the 1984–1986 period. Companies in which R&D-performing divisions had been sold off were deliberately excluded by NSF from its sample.

The results seemed quite striking.[9] In the 16 "merged" companies, R&D expenditure declined by 4.7 percent over 1986–1987, while the 8 "LBO/other restructuring" companies had R&D declining by 12.8 percent. The remaining 80 top-200 companies had seen R&D expenditures increase by 5.4 percent. While the NSF did not publish sales data which would allow one to look at research intensity in the different sets of firms, such data apparently will be made available shortly.

In thinking about these results, two points are worth bearing in mind. First, after inflation, the "normal" firms in the NSF sample had R&D growth rates which remained substantially below the pre-1986 norm (up 2

---

[8] My description here is based on various unpublished memos from the NSF.

[9] See National Science Foundation, "An Assessment of the Impact of Recent Leveraged Buyouts and Other Restructurings on Industrial Research and Development Expenditures," February 1, 1989.

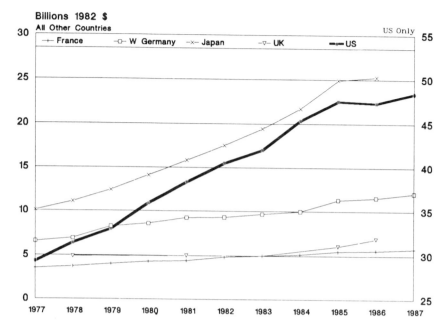

FIGURE 5   Industry funds for R&D.

percent after inflation). Thus, even if there is a link between restructuring and R&D, the slowdown in the nonrestructured firms remains to be
explained, and the attendant concerns persist.

Second, growth in foreign countries of inflation-adjusted company
R&D funds was not totally dissimilar to that in the United States. Figure 5
charts NSF data on industry funds for R&D (a slightly different definition
than that used earlier, since it includes funds for R&D performed outside
of the company itself) in the United States and five other industrialized
countries. It is interesting, and possibly important, to note that like the
United States, Japan and West Germany seem to have experienced a
slowdown in industry-funded R&D growth after 1985.[10] This of course
suggests that general business conditions in globalized high-technology
industries, rather than country-specific phenomena (like the U.S. LBO
wave) are the root cause of the slowdown.

At least two other large studies have attempted to look at the impact
of selected restructurings on corporate R&D investment in the 1980s, and

[10]These data are computed from numbers found in National Science Foundation, *International Science and Technology Data Update: 1988*, Special Report NSF 89-307, December 1988, pp. 4, 16. Foreign industry R&D funds have been converted to U.S. dollars at purchasing power parity exchange rates.

both studies contained at least some findings in apparent contradiction with the NSF results.[11] A study by Bronwyn Hall,[12] though primarily focused on models of corporate merger and acquisition behavior, tabulates data on the R&D intensity of merged firms one year before and after their merger, and compares it with that of firms not involved in merger activity within the same industry. Hall finds no obviously significant pattern of change in R&D intensity, either before and after merger, or between merged and nonmerging companies. The data cover the period 1976 to 1985 (which would cover mergers occurring in 1984), but Hall has stated that extending the sample to 1987 (covering mergers through 1986) leaves the conclusion unchanged.[13]

The other major set of studies considered here were undertaken by Frank R. Lichtenberg and Donald M. Siegel. The first Lichtenberg and Siegel study examined the impact of ownership changes on employment in firms' production and non-production establishments,[14] over the 1977 to 1982 period. As part of that analysis, they show that ownership changes over that period appeared to have no significant impact on R&D employment in companies' central offices. The second Lichtenberg and Siegel study directly tackles the issue of the impact of LBOs on company R&D expenditure by examining changes over time in the average research intensity of 43 firms involved in LBOs, and comparing the LBO firms' research-to-sales ratio with the average research intensity of all R&D-performing firms.[15] The

---

[11] We exclude from this discussion Kohlberg Kravis Roberts & Co., "Presentation on Leveraged Buy-Outs," January 1989, on the grounds that the data presented there—which include projections rather than actual historical numbers—are not comparable with the other studies discussed. See William F. Long and David J. Ravenscraft, "The Record of LBO Performance," July 1989. We also have omitted at least two other unpublished studies: Abbie Smith, "Corporate Ownership Structure and Performance: The Case of Management Buyouts," University of Chicago, January 1989; Michael A. Hitt, Robert E. Hoskisson, R. Duane Ireland, and Jeffery S. Harrison, "Acquisitive Growth Strategy and Relative R&D Intensity: The Effects of Leverage, Diversification, and Size," May 1989. Finally, F.M. Scherer and David J. Ravenscraft, *Mergers, Sell-offs, and Economic Efficiency* (Washington, D.C.: Brookings Institution), 1987, though analyzing mergers of the late 1960s and early 1970s which differed in some important respects from the buyouts of the 1980s, contains a detailed analysis of the impact of those mergers on R&D.

[12] Bronwyn H. Hall, "The Effect of Takeover Activity on Corporate Research and Development," in Alan Auerbach, Ed., *Corporate Takeovers: Causes and Consequences* (Chicago: University of Chicago Press), 1988.

[13] Bronwyn Hall, Testimony on "Corporate Restructuring and Its Effect on R&D," Hearings before the Science, Research, and Technology Subcommittee, House Committee on Science, Space, and Technology, July 13, 1989, p. 2.

[14] See Frank R. Lichtenberg and Donald Siegel, "The Effect of Takeovers on the Employment and Wages of Central-Office and Other Personnel," Bureau of the Census, Center for Economic Studies, Discussion Paper CES 89-3, June 1989.

[15] Frank R. Lichtenberg and Donald Siegel, "The Effects of Leveraged Buyouts on Productivity and Related Aspects of Firm Behavior," NBER Working Paper 3022, June 1989.

study found that average research intensity *increased* in the LBO firms over time, and even showed some increase relative to research intensity in all firms.

The second Lichtenberg-Siegel study is particularly interesting because it made use of data which included as a subset the same 200 largest R&D-performing United States firms used in the NSF study. While the NSF study used data on only 1 LBO in its "LBO/other restructuring" group (the remaining 7 firms fell in the "other" category, and largely were made up of firms undertaking stock buy-backs to defend against takeover bids), Lichtenberg found 43 firms undergoing LBOs within their sample. This difference is attributable to the much larger size of the Lichtenberg sample, the different definitions used to identify an LBO, the elimination from the NSF sample of companies with divisional sell-offs, and differing time periods over which the LBOs were counted (1978 to 1986, versus 1984 to 1986).

The simple fact is that LBOs as a group don't do much R&D. Note Bronwyn Hall's observation that the LBOs in her sample generally did much less R&D than the average manufacturing firm: "the recent increase in acquisition activity due to leveraged buyouts or other such private purchases is more or less orthogonal to the R&D activity in manufacturing. Even if all such purchases resulted in the complete cessation of R&D activity by the firm, this would amount to only around 500 million 1982 dollars annually compared to expenditures on R&D by the manufacturing sector of approximately 40 billion 1982 dollars annually."[16] Other information also points to the lack of much overlap between LBOs and R&D activity.[17]

How can these studies be squared with the seemingly contradictory NSF results? The first Lichtenberg-Siegel study is concerned with the earlier 1977–1982 period, and thus preceded the recent wave of leveraged takeovers. Furthermore, central office data on R&D employment include only central R&D laboratories, which accounted for a little under half of all industrial R&D employment in 1982. One can think of admittedly speculative scenarios where the engineering and research staff at central laboratories might be more insulated from declines in R&D spending than their brethren engineers in the field. When R&D budgets slow, R&D personnel stationed out in production and development facilities might be reassigned to other engineering tasks, while R&D directors retain staff at the central R&D facility.

---

[16]Hall, in Auerbach, Ed., 1988, p. 82.

[17]Hall, House Subcommittee on Science, Research, and Technology, July 13, 1989. The "LBO/other restructuring" group in the NSF study also did only a small amount of R&D in the aggregate.

Both the Hall and Lichtenberg-Siegel papers findings on research intensity (R&D-sales ratios) are not necessarily inconsistent with NSF's observation of an absolute decline in R&D in restructured firms. If sales fall as a consequence of an LBO, an increase in research intensity may accompany a decline in absolute levels of R&D. Since LBOs are thought to frequently be motivated by a desire to downsize a firm, to cut unprofitable lines of business and associated investments, this is not entirely unlikely.

In short, the NSF appears to have selected a particular group of firms—including both LBO/other restructurings, and more conventional mergers—in which absolute R&D spending in the mid-1980s rose much less than average. Hall and Lichtenberg-Siegel, looking at somewhat different kinds of restructurings in slightly different time periods, find little difference in research intensity between restructured and other firms. The apparent contradiction is reconcilable if sales declined at least as rapidly as R&D in the NSF's restructured firms. Whether this is the explanation will be known when the NSF releases aggregate sales data for the groups of firms in its sample.

Despite the ambiguity in the empirical evidence, opposing points of view on the consequences of LBO activity seem to concur in predicting diminished R&D activity in restructured firms. We next turn to a brief discussion of these opposing viewpoints.

## TWO POINTS OF VIEW ON LEVERAGED TAKEOVERS

The rise of the LBO in the 1980s attracted the attention of researchers, and led to explanations for this activity in terms of a "market for corporate control." This market is the aggregate of transactions undertaken by shareholders to create incentive and discipline mechanisms that will induce corporate managers to act in their interest. The empirical consensus seems to be that takeover activity generally increases the market value of the stock of companies involved.

One's view of whether this is socially desirable boils down to judgments about the source of that increased value. If the gain in market value results from the more efficient or productive utilization of a firm's assets or cash flows, then that ought to be counted in the benefit column and offset whatever costs involving R&D resources might be created. On the other hand, if the increase in the value of equity reflects a temporary misjudgment by the market, or comes at the expense of other groups with an interest in the firm—existing creditors who have seen their bonds or loans grow riskier without any compensating increase in return, or employees who have seen their wages and benefits drop when their contracts are renegotiated—then any negative impacts on R&D are not offset by gains for society as a whole.

## RESTRUCTURING AND R&D: A NEGATIVE VIEW

One pessimistic assessment of corporate restructuring's effects on research rests specifically on its assumed impact on the financial structure of the firm. The emphasis is on the "L" in LBO. Restructuring is assumed to lead to an increase in a corporation's debt-to-equity ratio, and a continuing need to pay out relatively large amounts in interest to service the increased debt. Interest payments on debt are fixed charges, which must be paid regardless of business conditions. R&D expenditure, on the other hand, can be deferred *in extremis*. Hence when sales are down, or costs are up more than anticipated, R&D—like other deferrable expenditures—is cut back. Increasing the debt load shouldered by the firm, it is argued, makes firms worry more about having enough cash flow to meet interest charges in times of adversity, and therefore concentrate more on short-term projects generating quick paybacks than would have been the case with a less leveraged financial structure.[18]

To put this argument in perspective, it is helpful to sketch out how a firm's optimal financial structure is determined.[19] Because interest payments are deductible from taxable income, and dividends are not, the tax system tips a company toward debt as a source of funds. However, in an uncertain world, increasing debt makes it more difficult to cope with unforeseen declines in earnings. In the worst case, the firm might be forced into bankruptcy. Because the process of bankruptcy is itself costly, the value of the firm's assets to creditors is reduced by the cost of their liquidation. Since the probability of bankruptcy depends on how leveraged the firm is, the real costs of bankruptcy make it desirable not to so increase leverage as to make bankruptcy highly likely.

In equilibrium, "the tax advantages of an extra dollar of debt would be just offset by the additional bankruptcy and moral hazard costs implied by the replacement of a dollar of equity with debt. Since the bankruptcy costs will vary by firm for many reasons, particularly the variability of the firm's earnings, the equilibrium debt-equity ratio will vary by firm, and the firm with more variable earnings will choose a lower debt-equity ratio."[20] (The

---

[18]This argument is laid out in Robert R. Miller, "The Impact of Merger and Acquisition Activity on Research and Development in U.S.-Based Companies," contractor report submitted to the Office of Technology Assessment, June 1989, referenced in Julie Fox Gorte, "Statement Before the Subcommittee on Science, Research, and Technology," House Committee on Science, Space, and Technology, July 13, 1989, p. 3.

[19]A clear and accessible exposition of the arguments sketched out here may be found in Roger H. Gordon and Burton G. Malkiel, "Corporation Finance," in Henry J. Aaron and Joseph A. Pechman, Eds., *How Taxes Affect Economic Behavior* (Washington, D.C.: The Brookings Institution), 1981.

[20]Gordon and Malkiel, p. 149.

moral hazard cost is the cost created by the possibility of shareholders or managers imposing uncompensated increases in risk on creditors holding debt instruments.)

Where does the impact of increased financial leverage on R&D fit into this framework? R&D investments are inherently highly uncertain, and a high-technology firm with many such projects and relatively volatile earnings would normally be expected to choose a lower debt-to-equity ratio. If a high debt-to-equity ratio were suddenly imposed on such a firm, one adjustment in moving back toward equilibrium would be to shed some of the riskier investment projects—like R&D, and particularly the most uncertain and long-term R&D projects—to reduce the variability of earnings. This is the essence of the argument made by critics of the impact of restructuring on R&D.

But is this cutting back of R&D in the restructured firm necessarily bad? If one firm dumps an R&D project that is viewed by the market as worthwhile, some other firm is free to invest in it. One firm's cutback in R&D, brought about by having a higher debt-to-equity ratio imposed, ought be matched by another firm's willingness to jump into the profitable vacuum if the project was truly worthwhile.

Of course, this may ignore other relevant costs. There may be firm-specific sunk costs associated with the start-up of an R&D project. These costs presumably must be duplicated if a terminated R&D project is started up elsewhere, and along with the opportunity costs incurred by delaying an innovation, represent a genuine social cost to the nation. And shuffling R&D resources among companies may itself involve real costs to society, costs divided among companies and researchers.

Critics of corporate restructuring also argue that other costs, unrelated to R&D, are imposed on society by excessive restructuring activity. Substantial resources may be absorbed by legal and financial maneuvers, with no real return to society.[21] The increased leverage created by restructurings may increase the business failure rate during a downturn, creating macroeconomic costs external to the firm. If the cash flows freed by cancelled investment projects are not reinvested productively (for example, if they are used to increase consumption), the future standard of living will fall. And some restructurings may increase the riskiness of existing debt, which lowers its value without compensating creditors for the increased risk. The moral hazard costs may create distortions in investment incentives, and a long-term efficiency cost to society.

---

[21] Some argue that the lawyers and financiers involved would otherwise be involved in other, equally unproductive activity, so that little net cost to society is created.

## BUYOUTS AND MANAGEMENT REFORM: THE POSITIVE VIEW

Another view focuses on the "h" in hostile takeover. From this perspective, the recent upsurge in hostile takeover activity—particularly leveraged buyouts—is beneficial to the United States economy. This view takes as its point of departure the so-called "free cash flow" theory of Michael Jensen.[22] Briefly, Jensen notes that corporate managers are supposed to function as the agents of shareholders, implementing shareholder interests in maximizing the value of the firm. But conflicts of interest between managers and stockholders may arise. "Agency costs"—the costs of monitoring managers and devising incentives for them to act in the interest of stockholders, and the losses inflicted on stockholders by such conflicts of interest—may be great. The possibility of a buyout or hostile takeover, from this point of view, may serve as a discipline and deterrent for managers who get too far out of line with the interests of stockholders. Such corporate restructuring, or its threat, serves the positive social function of reducing agency costs.

Jensen argues that some firms may face the problem of free cash flow, "cash flow in excess of that required to fund all of a firm's projects that have positive net present values when discounted at the relevant cost of capital. Such free cash flow must be paid out to shareholders if the firm is to be efficient and to maximize value for shareholders."[23] Managers, on the other hand, may prefer to use those resources in ways designed to increase their influence and compensation, by plowing funds into investment projects in order to increase the company's size or growth rate. It has been argued that such free cash flow—and hostile takeover activity—should increase when high real interest rates make many projects less attractive investment opportunities, or as industries mature and the most profitable projects are exhausted.[24] Increased debt, in addition to making the restructuring possible, creates a fixed charge against future cash flows. If debt is exchanged for equity, argues Jensen, this has the effect of forcing managers to promise increased future cash flows to stockholders.

R&D investments are presumably among some of the insufficiently profitable projects that managers may be pursuing at stockholder expense. Therefore, one consequence of corporate restructuring designed to overthrow incumbent management and divert free cash flow to stockholders would be to expect cutbacks in R&D investment projects. However, this

---

[22] See Michael C. Jensen, "Takeovers: Their Causes and Consequences," *Journal of Economic Perspectives*, vol. 2, no. 1, Winter 1988.

[23] Jensen, 1988, p. 28.

[24] See Margaret M. Blair and Robert E. Litan, "Explaining Corporate Leverage and LBO Activity in the Eighties," mimeo (Washington, D.C.: The Brookings Institution), 1989.

cutback in R&D would be viewed *positively* by defenders of the takeover, since the resources would be shifted to more profitable activities.

One important point made by Jensen is that the *example* created by hostile takeovers against unprofitable investment of free cash flow by management serves to discipline other managers observing these events. The hostile takeovers of some firms may induce other firms not subject to an actual takeover attempt to respect stockholder interests, and cut insufficiently profitable activities.

## SOME TENTATIVE ASSESSMENTS

Because technology investments are central to our future standard of living, the impact of corporate restructurings on R&D is an important issue. More work on the interactions between distinct types of mergers, sell-offs, and research and development activity is clearly required before we can have any real confidence in our understanding of these linkages. To date, empirical work on the response of R&D investment to recent corporate restructurings has largely been confined to comparing investment ratios among broad groups of firms.

At least one set of arguments suggests that the crucial relationship is between financial leverage (which may vary significantly among merged firms) and R&D investment. For free-cash-flow theorists, it is the hostility of the takeover which is the key to understanding its effects. Both perspectives suggest that different types of mergers and acquisitions should be distinguished from one another for the purposes of analysis. Another set of arguments suggests that some significant effects of restructuring are felt outside the acquired firm. These issues complicate interpretation of existing work, and merit further empirical study.

Nonetheless, a few preliminary judgments may be ventured. First, there is little evidence for the proposition that "restructuring" reduces research intensity in affected firms.

Second, if LBOs generally lead to reductions in sales by acquired firms, it is entirely possible that declines in the absolute level of R&D spending might occur despite constant or increasing R&D intensity.

Third, R&D activity on the part of firms taken over in recent LBOs is probably too limited to have had much more than a marginal direct impact on aggregate industrial R&D. The possible indirect impact on the behavior of other companies remains unexplored.

The similarity of company R&D spending trends abroad to those in the United States suggests other influences may be more important. Even among companies undergoing no restructuring, R&D spending slowed significantly. Those concerned with the recent decline in company-funded

R&D growth would be well advised to investigate other explanations as well.

Finally, the ultimate test of the impact of the recent LBO wave has yet to be faced. That will come when the United States economy enters a significant recession.

# Appendix B
# List of Participants

Thomas Althuis
Director, Science Policy Affairs
Pfizer, Inc.

Roger C. Altman
Vice Chairman
The Blackstone Group

Howard Banks
Bureau Chief
*Forbes*

Claude Barfield
Senior Fellow
American Enterprise Institute

Theodore M. Bednarski
Vice President, Science & Technology
Hercules Incorporated

Margaret Blair
Research Associate
Brookings Institution

Some participants did not attend all sessions.

Jacques A. Bodelle
Representative for the United States
Corporate Research & Innovation
Elf Aquitaine

I. MacAllister Booth
President and Chief Executive Officer
Polaroid Corporation

Charles S. Bridge
Vice President and Chief Scientist
Litton Industries, Inc.

John G. Brim
Managing Director
Merrill Lynch

Harry Broadman
Chief Economist
Senate Governmental Affairs Committee

Daniel Burton
Vice President
Council on Competitiveness

John W. Collette
Assistant to the Vice President, Central Research
E. I. du Pont de Nemours and Company

William J. Cook
Senior Editor
*U.S. News & World Report*

Douglas P. Crain
Director, Aircraft Products Group
LTV Aerospace and Defense Co.

Richard L. Crawford
Director, Federal Affairs
National Restaurant Association

Michael Davey
Analyst in Science & Technology
Congressional Research Service

Robert Davis
*Wall Street Journal*

Richard R. Dickinson
Vice President
Texaco Inc.

Robert Dillon
Executive Vice President
Sony Corporation of America

John O. Dimmock
Staff Vice President
McDonnell Douglas Research Laboratories

Gerald P. Dinneen
Foreign Secretary
National Academy of Engineering

James Ebert
Vice President
National Academy of Sciences

Stuart E. Eizenstat
Partner
Powell, Goldstein, Frazer & Murphy

Richard Endres
Deputy Assistant Secretary
Office of Technology Policy
Department of Commerce

Kevin Finneran
Senior Editor
*Issues in Science and Technology*

Kenneth S. Flamm
Senior Fellow
The Brookings Institution

Alexander H. Flax
Home Secretary
National Academy of Engineering

Philip H. Francis
Vice President, Corporate Technology Center
Square D Company

Glen S. Fukushima
Deputy Assistant U.S. Trade Representative for Japan and China
Office of the U.S. Trade Representative

Stuart Gannes
Associate Editor
*Fortune*

Thomas F. Gannon
Director, Technology Planning and Development
Digital Equipment Corporation

Richard A. Gephardt
Majority Leader
U.S. House of Representatives

Robert E. Grady
Associate Director
Office of Management and Budget

Edward Greelegs
Director of Congressional Relations
Federal National Mortgage Association

Carol Grundfest
Assistant Vice President, Science and Technology
Pharmaceutical Manufacturers Association

Joseph A. Grundfest
Commissioner
Securities and Exchange Commission

Bronwyn Hall
Department of Economics
University of California, Berkeley

Philip W. Hamilton
Managing Director, Public Affairs
American Society of Mechanical Engineers

Steve Hardis
Vice Chairman and Chief Financial Officer
Eaton Corporation

Dr. William M. Haynes
Director, Physical Sciences Center and External R&D Funding
Monsanto Company

Zachary Hernandez
Staff Assistant to Commissioner Grundfest
Securities and Exchange Commission

Christopher T. Hill
Senior Specialist, Science and Technology
Congressional Research Service

J. French Hill
Deputy Assistant Secretary for Corporate Finance
Department of the Treasury

Max Holland
Contributing Editor
*The Nation*

Kent Hughes
Democratic Policy Committee
U.S. Senate

Nathan Hurt
Manager, Business Development
Los Alamos Technical Associates

Len F. Impellizzeri
Staff Vice President, Engineering and Advanced Systems
McDonnell Douglas Corporation

George T. Jacobi
Vice President, Technology
Johnson Controls Inc.

Pat Janowski
*News Report*
National Research Council

Gregg A. Jarrell
Graduate School of Business Administration
University of Rochester

Denis King
Committee on Science, Space and Technology
U.S. House of Representatives

Peter O. Kliem
Senior Vice President and Director of Research
Polaroid Corporation

Richard Krashevski
Economist
General Accounting Office

Ralph Landau
Center for Economics Policy Research
Stanford University

Charles F. Larson
Executive Director
Industrial Research Institute, Inc.

Wil Lepkowski
*Chemical & Engineering News*

David Lewin
Media Relations Manager
American Society of Mechanical Engineers

Frank Lichtenberg
Graduate School of Business
Columbia University

Mark Lieberman
Director, Office of Technology Commercialization
Department of Commerce

William Long
Guest Scholar
The Brookings Institution

Walter P. Lukens
Vice President
Litton Industries

Roger Majak
Manager, Federal Government Affairs
Tektronix, Inc.

Gene G. Mannella
Director
Electric Power Research Institute

John Marano
Vice President
Olin Corporation

William A. Maxwell
Vice President
National Association of Manufacturers

Dan P. McCurdy
Program Manager
IBM Corporation

James A. McDivitt
Senior Vice President
Rockwell International

Linda McDonough
Senior Policy Analyst
Department of the Treasury

Richard A. Meserve
Partner
Covington & Burling

Larry Meshel
Research Director
Economic Policy Institute

Robert Miller
Department of Economics
University of Houston

William Morin
Director, Council of High Technology
National Association of Manufacturers

Deborah Morman
Staff Associate
Merck & Company, Inc.

Dennis Mueller
Department of Economics
University of Maryland

Stephen D. Nelson
Manager, Science Policy Studies
American Association for the Advancement of Science

Frank V. Nolfi, Jr.
Vice President, Technology
Alcan Aluminum Corporation

Colin Norman
*Science*

Henry Owen
Principal
Consultants International Group

Harold W. Paxton
U.S. Steel Professor
Carnegie Mellon University

Albert C. Perrino
Vice President, Technology and Strategy
ICI Americas Inc.

Helena Hutton Peterson
Government Relations Manager
3M

Marshall Phelps
Director, Governmental Programs
IBM Corporation

Jonathan Piel
Editor
*Scientific American*

John Pilcher
Director of Corporate Finance
National Association of Manufacturers

Melissa Pollak
Science Resources Analyst
National Science Foundation

Penelope Pollard
Senior Analyst
Office of Technology Assessment

William R. Prindle
Associate Director, Research, Development & Engineering
Corning Incorporated

Thomas Ratchford
Associate Executive Director
American Association for the Advancement of Science

Melissa Rhodes
Staff Assistant to Commissioner Grundfest
Securities and Exchange Commission

William Risen
Department of Chemistry
Brown University

Dorothy Robyn
Senior Analyst
Office of Technology Assessment

Robert Rosenblum
Legal Counsel to Commissioner Grundfest
Securities and Exchange Commission

Richard L. Rosenthal
Chairman
The Richard and Hinda Rosenthal Foundation

Ian M. Ross
President
AT&T Bell Laboratories

Ellis Rubinstein
News Editor
*Science*

Robert Samuelson
Columnist
*Newsweek*

Wendy Schacht
Specialist, Science & Technology
Congressional Research Service

J. Clifford Schoep
Corporate Director
General Dynamics Corporation

Elliott Schwartz
Chief, Commerce Unit
Congressional Budget Office

Robert Shapiro
Vice President
Progressive Policy Institute

Robert C. Sigrist
President, Defense Division
Brunswick Corporation

William J. Stape
Director, Optoelectronics
AMP Incorporated

Robert L. Stern
Consultant
RLS Associates

H. Guyford Stever
Former Foreign Secretary
National Academy of Engineering

William Stewart
Director, Division of Science and Resources Studies
National Science Foundation

Robert Stratton
Vice President and Director, Central Research Laboratories
Texas Instruments Incorporated

Michael Telson
Energy & Science Analyst, Budget Committee
U.S. House of Representatives

Michael Tokarz
Associate
KKR and Company

George Tyler
Senior Economist
Joint Economic Committee

Guilio Vita
Senior Vice President
Bristol-Myers Company

David Walters
Chief Economist
Office of the U.S. Trade Representative

Alfred J. Watkins
Economist, Strategic Planning Division
World Bank

Philip Webre
Principal Analyst
Congressional Budget Office

Leonard Weiss
Staff Director
U.S. Senate Committee on Governmental Affairs

Henry Wendt
Chairman
SmithKline Beecham

Erwin Whitman
Vice President—Medical Affairs
Bristol-Myers Company

Deborah Winegarden
Medill News Service

Deborah Wince-Smith
Assistant Secretary for Technology Policy
Department of Commerce

Michael Wolff
Editor
*Research Technology Management*

Sam Yanes
Director, Corporate Communications
Polaroid Corporation

Karl H. Zaininger
President
Siemens Corporate Research Inc.

*Staff: National Academy Complex\**

Frank Press
President
National Academy of Sciences

Robert M. White
President
National Academy of Engineering

---

\*The term "National Academy Complex" includes the National Academy of Sciences, the National Academy of Engineering, the Institute of Medicine, and the National Research Council.

Edward Abrahams
Senior Staff Officer
Academy Industry Program
National Research Council

Deborah Faison
Senior Program Assistant
Academy Industry Program
National Research Council

Alan Fechter
Executive Director
Office of Scientific and Engineering Personnel
National Research Council

Bruce Guile
Director, Program Office
National Academy of Engineering

Kathi Hand
Research Assistant
Office of Government Affairs
National Research Council

Nancy Gardner Hargrave
Director, Office of Development
National Academy of Sciences

Allan R. Hoffman
Executive Director
Office of Government & External Affairs
National Research Council

Scott Lubeck
Director
National Academy Press

Angie Maurolis
Office of the President
National Academy of Sciences

Lawrence McCray
Executive Director
Committee on Science, Engineering, and Public Policy
National Academy of Sciences

Stephen A. Merrill
Director, Office of Government Affairs
National Research Council

Norman Metzger
Deputy Executive Officer
National Research Council

Proctor Reid
Fellow
National Academy of Engineering

William C. Salmon
Executive Officer
National Academy of Engineering

Kenneth Smith
Senior Secretary and Project Assistant
Academy Industry Program
National Research Council

Philip M. Smith
Executive Officer
National Academy of Sciences/National Research Council

Myron F. Uman
Acting Executive Director
Commission on Physical Sciences, Mathematics, and Resources
National Research Council

John S. Wilson
Study Director
Committee on Science, Engineering, and Public Policy
National Academy of Sciences

# Appendix C
# Agenda

**OCTOBER 11, 1989**

6:30 p.m.    Reception

7:30 p.m.    Dinner

8:30 p.m.    **Welcome**            Frank Press, President,
                                     National Academy of Sciences

             **Keynote Address**    Joseph A. Grundfest, Commissioner,
                                     Securities and Exchange Commission

**OCTOBER 12, 1989**

7:30 a.m.    Continental Breakfast

8:30 a.m.    **Welcome**            Robert M. White, President,
                                     National Academy of Engineering

8:45 a.m.    **Introduction**       Stuart E. Eizenstat, Partner,
                                     Powell, Goldstein, Frazier and
                                     Murphy

| | | |
|---|---|---|
| 9:00 a.m. | **Wall Street View** | Roger C. Altman, Vice Chairman<br>The Blackstone Group |
| | | Michael T. Tokarz, Associate<br>KKR & Company |
| 10:30 a.m. | Break | |
| 10:45 a.m. | **Corporate View** | I. MacAllister Booth, President and<br>  Chief Executive Officer<br>Polaroid Corporation |
| | | Henry Wendt, Chairman<br>SmithKline Beecham |
| 12:15 p.m. | Lunch | |
| 1:00 p.m. | **Public Policy** | Richard A. Gephardt, Majority<br>  Leader<br>U.S. House of Representatives |
| 2:00 p.m. | **Overview** | Kenneth S. Flamm, Senior Fellow<br>The Brookings Institution |
| | | Gregg A. Jarrell, Professor<br>University of Rochester |
| | | Dennis C. Mueller, Professor<br>University of Maryland |
| 3:45 p.m. | **Summary** | Stuart E. Eizenstat |
| 4:00 p.m. | Adjournment | |